FANTASTIC READING

Stories
and
Activities
for
Grades 5-8

FANTASTIC READING

Stories and Activities for Grades 5-8

Isaac Asimov

Martin H. Greenberg

David Clark Yeager

Scott, Foresman and Company
Glenview, Illinois
London

Dedication

To Professor Harry Kane, who always
inspires the love of learning

ISBN: 0-673-15936-1

2 3 4 5 6 7 - PAT - 89 88 87

FOREWORD
Isaac Asimov

Modern man is, perhaps, 50,000 years old. For 90 percent of that time, every human being who ever lived was illiterate, as it was only 5,000 years ago that writing was invented. Without such a system for the permanent preservation of thought, civilization could not have advanced much further.

Ever since the ancient Sumerians began to write, the technological and intellectual advances of humanity have depended upon the ability to convert speech easily into a code of markings and to reconvert that code rapidly into speech at sight. In other words, people had to be able to write and then to read what was written.

Yet, during 97 percent of the time that the knowledge of writing existed in the more advanced societies, it remained the private possession of a small minority of the population. The priestly caste could read and write. Merchants might. Warriors didn't need to and frequently scorned literate persons as feeble effeminates. Rulers and aristocrats were sometimes illiterate, but they could easily hire scribes to do the job for them. Peasants and women, who together constituted 90 percent or more of the total adult population, were considered as having no use whatsoever for literacy.

Nevertheless, as technology advanced and as civilization grew into larger and more complex units, the ability to read and write became more essential. The turning point came with the Industrial Revolution in Great Britain toward the end of the 17th century.

If farm laborers could afford to be illiterate, factory workers could not. Because the new industries had complex machines, instruction — and communication generally — could not be carried on by speech alone. It is no accident that as various countries became industrialized, education underwent a revolution of its own. Mass education became a necessity and, for the first time in human history, societies developed in which most individuals could read and write.

But although technology has continued to move ahead, education has not kept pace. Our schools are not educating children properly in the sciences and mathematics, areas of study that will be essential for those who must live in the high-tech society that is now rapidly inundating us. However, unless we first make certain that our children can read and write, nothing will help us teach science, mathematics, or anything else that is needful. One cannot teach anything to illiterates that involves even a modicum of intellectual complexity.

But is literacy not almost universal in the United States? A great deal depends on how we define literacy. If we are satisfied to consider anyone who can write his or her name in block letters or who can name the letters on a stop sign, then, indeed, almost all Americans above the age of seven are literate.

Suppose, though, we decide that a literate person must be able to read and understand, with reasonable ease, a hundred words of small print; to be able to fill out a job application; to be able to read and follow instructions as to what to do in case of fire; to be able to understand a description of how to operate a simple machine. In that case, many Americans cannot be considered literate. They may be able to make out the headlines on the sports page, but they are functionally illiterate.

Is such illiteracy unavoidable? Or must our schools somehow improve their methods of instruction and find ways to stimulate that portion of our young people who, for one reason or another, seem reluctant to apply themselves to the task of learning the code? That is, find ways of supplying the motivation that doesn't seem to arise spontaneously from within in these cases.

It is this problem that prompted my involvement in *Fantastic Reading*. I think the book helps fill a desperate need, for the device it uses is a good one.

We must find materials that children are eager to read — something that they will enjoy, something that will reward them for what they might consider a tedious effort. I firmly believe that nothing fulfills this purpose as well as science fiction and fantasy.

There are a number of reasons for this. In the first place, young minds lend themselves to expanded horizons. They have not yet walled themselves in with the prosaic or been beaten down by the every-day. They not only welcome the strange and the outré, they are excited and delighted by it.

Secondly, we all *live* in a world of science fiction. Those of us who are old enough to recall World War II remember that in the early 1940s there were no nuclear weapons, no jet planes, no television, no rocket ships, no computers, no lasers. Yet now human beings have landed on the moon, and our space probes have taken close-up pictures of the rings of Saturn. We have robots on the assembly lines and missiles that can destroy the world. I myself wrote science fiction stories about such things before they existed, and "practical" people laughed at me. Now people regard me with awe and consider me remarkable. (Actually, I'm neither laughable nor awe-inspiring. I'm just a science fiction writer.)

The changes that have occurred in the last 40 years have been so radical and so rapid that the general population has at least become convinced of the inevitability of change. Even youngsters understand this, and they also understand that the world they will live in as adults will be far different (whether for better or worse) than the world they live in now. Science fiction and fantasy symbolize strange worlds. Although they may not describe the *particular* strange worlds that present-day children will someday inhabit, they are strange, and children gravitate naturally toward them.

Even though science fiction and fantasy in their printed form are still a little daunting, an increasing number of such books are making the best-seller lists (including, to my vast surprise, two of my own). And when science fiction and fantasy are made even more accessible by presenting them in the form of movies with elaborate special effects, one achieves the runaway popularity of such motion pictures as *Star Wars* and *E.T.* But why leave science fiction and fantasy only to entertainment? Why not use them in education for the reading needs of the nation's children and to promote their interest in science?

There's an additional, more specialized point that I might make, one that I can draw both from my own experience and from the letters I receive from my readers. When I was a child, there were no books in our home because my parents could not afford to buy any. My father owned a small store, however, in which, among other things, he sold magazines. But he was reluctant to allow me to read them because he had decided, after sampling them himself, that they were too violent and trashy for me. However, he made an exception in the case of science fiction magazines, as he thought these might prove educational.

He was right. I read my first science fiction stories at the age of nine and was instantly fascinated. Most of the stories I read were probably just as "trashy" as those in the magazines that were forbidden to me, but at least some of them dealt with scientists and with science in a dramatic and literate way, and I date my life-long interest in science from that time.

I began to find science books in the public library and to read all I could about the subject. The result was that I majored in chemistry when I went to college, and now I am Professor of Biochemistry at Boston University School of Medicine (even though I no longer work at it actively).

Mine is not an isolated case. For 45 years I have been writing science fiction, and in that time I have received thousands upon thousands of letters from readers. A surprising (and gratifying) number of them tell me that my science fiction has led them on into one branch of science or another.

It is my belief, then, that science fiction is not only ideally suited to encourage youngsters into *wanting* to read, it also may help influence a number of them to participate in the future of advanced technology — to their benefit and ours.

CONTENTS

INTRODUCTION

Fantastic Reading contains two elements essential to successful instruction: academically appropriate material and a clear format. The readings and activities in this book can be used effectively in a variety of learning environments. Featuring short selections by respected science fiction authors, *Fantastic Reading* generates strong reader interest while also developing specific language skills. Accompanying each selection are activities that support reading, vocabulary, writing, and content skills. All of these activities include concise student instructions, allowing the material to be used effectively with large groups as well as with the independent learner. Because none of the stories contains materials requiring any special background knowledge, *Fantastic Reading* is a practical educational tool.

HOW TO USE THIS BOOK

The structure of *Fantastic Reading* encourages complete flexibility in the development of a total-language program. The reading selections, with their supporting activities, are designed to be individually reproduced for maximum convenience and flexibility. They can be used in any order selected by the instructor. The amount of time necessary to complete an entire unit of study can be adjusted to meet the demands of varying class lengths and schedules.

Contents, Skills Matrix, and Index

The subject divisions found in the Contents arrange the stories under three broad topics: Creatures, Macabre, and Fantasy. The Skills Matrix at the end of this Introduction and the Index at the back of the book may be used to select readings keyed to a specific language or comprehension skill. These references also simplify record-keeping associated with individualized learning programs.

Reading Selections

The seventeen short stories in *Fantastic Reading* have enjoyed wide success. They are not abridged versions of longer material. Students should have little difficulty reading an entire selection in one session.

Vocabulary Development Exercises

Vocabulary words that may prove difficult for some students are introduced before each story, in the order in which they appear in the story. The descriptive sentence following each definition refers to a character or event in the story. Thus, the purpose of the Prereading Vocabulary is to increase reading comprehension without revealing the plot. The vocabulary should be presented on the board, on a duplicated handout, or on an overhead transparency. Another alternative is to introduce the vocabulary from several of the reading selections on a bulletin board or on individual index cards.

The Postreading Vocabulary Exercise following each story challenges the students to broaden their understanding of language. One clear sign of student proficiency is the ability to transfer information from one context to another. Many of the activities simulate the format found on many standardized tests, providing valuable test-taking practice. Other activities provide a recreational approach to vocabulary development.

Discussion Questions

The discussion questions, "Thinking About What You've Read," that follow each reading will assist in making an informal assessment of students' comprehension. Further, these questions help students clarify their recall of pertinent information and develop opinions about the material they have just read. These questions will be effective whether they are presented orally, drawing responses from group interaction, or in written form, providing individual interpretations of the story.

Reading Comprehension Exercises

Two reading comprehension exercises accompany each reading. The scope of these activities is consistent with the reading curricula of most state and local school districts. Specific skills are listed in the Skills Matrix at the end of this Introduction. Each comprehension assignment is preceded by clear and concise instructions to the student. Examples of appropriate responses have been included where necessary. Since none of the activities requires access to specific reference materials, exercises may be given as homework assignments.

Content and Writing Skills

Academic achievement is dependent not only on the quality of instruction, but also on the ability of the student to research, organize, and use knowledge. The content skills activities promote competency in a variety of academic, consumer, and organization skills. The students, upon completion of a content skills exercise, should then use the information they have developed as the structure of their writing activity.

Admittedly, providing the structure for an effective writing assignment is difficult to do. A complex set of instructions may limit student creativity. Writing assignments with little or no structure, however, are often too broad to promote student success. But by using the information from the previous activity, the major obstacles to providing an appropriate structure for students are removed. Students are able to use their own work as the prewriting motivation necessary for an exemplary finished product.

SAMPLE SEQUENCES

The following sample sequences suggest how the materials in this book might be used in either group instruction or in an independent learning program. Activities designated by "(H)" in the sequences for group instruction will serve as suitable language arts homework assignments.

Group Instruction

Session 1

A. Introduce story.
B. Distribute Prereading Vocabulary and discuss.
C. Motivate reading through discussion of characters and setting.
D. Distribute and read selection.

Session 2

A. Review story using discussion questions.
B. (H) Assign Postreading Vocabulary Exercise.
C. (H) Assign comprehension activities (one of the activities might be completed orally with a group).

Session 3

A. Distribute both the content and the writing activities.
B. Preview instructions; assign content activity.
C. Evaluate student performance on content activity.
D. (H) Encourage use of the content activity as a motivation for the writing activity; assign writing activity.

Independent Learning Program

Session 1

A. Student selects reading unit.
B. Student completes vocabulary preview and checks responses.
C. Student reads selection.

Session 2

A. Student completes Postreading Vocabulary Exercise.
B. Student reviews story by responding to one or more of the discussion questions.
C. Student selects one or both of the comprehension activities, checking responses when finished.

Session 3

A. Student completes the content skills assignment.
B. Responses on the content activity are evaluated and necessary corrections made.
C. Student begins the final writing assignment.
D. A unit conference between instructor and student is held when all assigned activities are completed.

COMPREHENSION SKILLS MATRIX

PART ONE

CREATURES

INTRODUCTION TO PART ONE

Isaac Asimov

Any good story should raise more questions than it answers. A story that too neatly ties together all loose ends is somehow unsatisfying, as it leaves nothing to talk and wonder about. A fairy tale that ends with "and they all lived happily ever after" may give young people a feeling of security, but surely there comes an age when one would rather ask, "And what happens next, I wonder?"

What better can a story do than give you something to think about? That means the story isn't over when it seems to be over; rather, it leaves behind something that will keep you thinking. When you're dealing with a very short story, this is particularly true. Because the writer doesn't have time to go far beyond the immediate point in one or two thousand words, there is much he or she is forced to leave to the readers.

In "How Now Purple Cow," students might argue intensely as to what the *purpose* of the purple cow might be. Is it possible that as more and more people touched one or the other purple cow, they too might become purple cows? Would everyone on Earth become a purple cow? Would people begin to realize what was happening after several changes and then kill and burn all the strange creatures? Would the purple cows be able to defend themselves against this? It shouldn't take long for the students to see that one could write an entire novel on this theme rather than just a short-short story. A number of them might even want to try their hand at continuing the story; obviously, they should be encouraged to do so.

Sometimes a story is written lightly, and it should be read with that point in mind. "Just Call Me Irish" is written with deliberate humor, and students should be encouraged to note that reading can sometimes make you laugh. The more subtle the humor, the more delightful it is. When Irish speaks of "dogged persistence" and complains of "dogma," have students explain why such remarks might make readers chuckle. (Because even professional psychologists have trouble explaining humor, it should not be surprising if the students' explanations are many and varied.)

A good, light-hearted story should, however, also lend itself to possible serious discussion. Suppose dogs *were* intelligent. Does that mean houses should be built for every dog family? How many houses would have to be built? Would dogs want the same kind of houses that people live in?

"Dog Star," another dog story, can induce altogether different thoughts. Surely the Sirians are going to find out their mistake someday, and when they do, what will happen? Here, incidentally, is an example of the value of titles. Why is the story called "Dog Star"? That question might stimulate discussions of constellations that could prove just as interesting as the story itself.

A sudden inversion of situations that are usually taken for granted can lead to speculations far beyond the immediate story. "Zoo" is bound to give rise to thoughts concerning the ordinary zoos that students have visited. Is it possible that monkeys think *we* are funny?

There are many national parks and animal refuges in which wild animals are allowed to roam freely, while people are cautioned to remain in their cars for their safety. Of course, we don't stay in such places very long, but suppose we *had* to stay there and could *never* get out of our cars. Would we like it? Well, then, do animals like to be always behind bars?

A reading of "Zoo" might encourage students to think about the humane treatment of zoo animals, which is something that might not ordinarily occur to them. Such considerations are of greater value when students think of them spontaneously as a result of reading a story that, on the face of it, does not bring up the matter.

Finally, a story such as "Creature of the Snows" can introduce philosophical points that pose dilemmas even professional scientists find difficult to resolve. Yet students might find themselves so interested in such problems that they don't hesitate in taking firm positions. For example, it is one thing to ask simply, "Do yetis exist?" That, in itself, can create controversy. But suppose you ask instead, "What happens if there is evidence for the existence of yetis that people simply refuse to believe? Is it better just to keep quiet?" Now we have a puzzling problem. But, then, the more students are puzzled, the harder they must think, and we want them to think hard.

How about the question, "What happens if by telling the truth you can endanger the welfare of creatures who are doing no harm and who may be wiped out if they are discovered?" An informal debate could easily be organized on such points, and the story can be re-read for evidence that the author is taking one side or another.

HOW NOW, PURPLE COW

Bill Pronzini

Prereading Vocabulary

grazing: feeding on growing grasses. (*Floyd first spotted the purple cow* grazing *near the farm.*)

hallucination: an unreal view of objects or events; a delusion usually caused by a mental disorder or as a response to a drug. (*When Floyd saw the purple cow, he thought he was having an* hallucination.)

chlorophyll: the green coloring in plants necessary for them to produce carbohydrates (for a detailed explanation of chlorophyll, see an encyclopedia). (*The strange color of the cow contrasted with the* chlorophyll *green of the grass.*)

foreman: the person in charge of a group of workers on a farm or in a factory; the boss. (*Floyd could not convince his* foreman *to come to the south pasture to see the cow.*)

corrugated: shaped into a series of alternating ridges and grooves. (*When Floyd wrinkled his forehead, it reminded his wife of a* corrugated *box.*)

oblivious: unable to recall; unaware. (*The cow seemed* oblivious *to Floyd's presence.*)

When Floyd Anselmo saw the purple cow grazing on a hillside of his dairy ranch one cold morning in October, he thought his mind must be hallucinating.

He brought his pick-up truck to a sharp halt at the side of the access road that wound through his property, set the brake, and leaned across the seat to have another look. But it was still there. He stared, willing it to disappear. It didn't.

Anselmo shook his head slowly and got out of the truck. He stood on the graveled roadbed, shading his eyes from the glare of the winter sun. Still there.

By God, Anselmo thought. Next thing you know, it'll be pink elephants. And me not even a drinking man.

He drew the collar of his coat up against the chill, early morning wind, sighed deeply, and walked around the truck. He made his way carefully through the damp grass at the side of the road, climbed easily over the white fence there, and began to ascend the hillside.

Halfway up, he paused for another look. Darned if the cow *wasn't* purple; a rather pleasant, almost lilac, shade of that color. Still, the contrast with the bright chlorophyll green of the grass, and the dull, brown-and-white of the other cows, was rather startling.

Anselmo climbed to within twenty feet of where the purple cow was grazing. Cautiously, he made a wide circle around the animal. It paid no attention to him.

3

"Listen here," he said aloud, "you ain't real."

The cow chewed peacefully, ignoring him.

"Cows ain't purple," Anselmo said.

The animal flicked its tail lightly.

He stood looking at it for quite some time. Then he sighed again, rather resignedly this time, turned and started down the hillside.

His wife was finishing the breakfast dishes in the kitchen when he came in a few minutes later. "Back so soon?" she asked.

"Amy," Anselmo said, "there's a purple cow grazing on a hillside down the road."

She wiped her hands on a dish towel. "I made some fresh coffee," she told him.

Anselmo tugged at his ear. "I said, there's a purple cow grazing on a hillside down the road."

"Yes, dear," his wife said. She began stacking dishes in the cupboard.

Anselmo went outside. He saw Hank Raiford, his foreman, coming up from the milking barn.

"Morning, Mr. Anselmo," Hank said.

"Hank," Anselmo said, frowning, "I saw this purple cow grazing on a hillside down the road."

Hank looked at him.

"I thought it was an hallucination at first. But I went up there and the damned thing was purple, all right. I can't figure it out."

"Well," Hank said, watching him strangely.

"You haven't seen it by any chance, have you?"

"No, sir."

Anselmo nodded. "Want to come out with me and have a look at it?"

"Well," Hank said, "there are a few things I got to take care of right now."

"Maybe later," Anselmo said.

"Sure," Hank told him, moving away quickly. "Maybe later."

Anselmo went back into the house. He crossed directly to the telephone on the hall table and put in a call to Jim Player, the editor of the local weekly newspaper.

"Floyd Anselmo here," he said when Player came on.

"What can I do for you, Floyd?"

"Well," Anselmo said. "I was coming into town a while ago, and I was driving down my access road when I saw this purple cow grazing on a hillside."

There was silence from the other end.

"Jim?" Anselmo asked.

"Purple cow?" Player said finally.

"That's right," Anselmo told him. "Purple cow."

Another silence, shorter this time. Then Player laughed. "You're putting me on, right?"

"No," Anselmo said seriously.

"Look, Floyd, I'm a busy man," Player said. "With all these silly damned UFO sightings hereabout lately, I haven't had time to..." He broke off, chuckling. "Say, maybe this purple cow of yours came in one of those flying saucers people claim to have been seeing."

"Jim," Anselmo said slowly, "I don't know anything about flying saucers. All I know is there's a purple cow grazing on one of my hillsides. If you want to come out here, I'll show it to you."

Player was silent for a moment. Then he said, "All right, I'll come out. But if you're ribbing me..."

"The hillside I'm talking about is maybe a mile onto my land from the highway," Anselmo told him. "I'll meet you there."

"Forty-five minutes," Player said unhappily, and hung up.

Anselmo went to the door. His wife came into the room just as he reached it. "Where are you going, dear?"

"To meet Jim Player."

"Whatever for?"

"To show him the purple cow I saw."

Her forehead corrugated worriedly. "Floyd..."

"I'll be back in an hour or so," Anselmo said, and stepped outside.

He started his pick-up and drove down the access road. When he reached the hillside, he saw that the purple cow had moved farther down it, and was grazing now only a few feet from the white fence.

Anselmo braked the truck and got out. He went through the grass to the fence, climbed over it, and stood facing the cow.

The animal continued to graze, seemingly oblivious to his presence.

Anselmo walked haltingly up to it. He put

out a wary hand and touched its head. Then he stepped back. "I was beginning to have some doubts," he said, "but damned if you ain't real, and damned if you ain't purple."

The animal shifted its hind legs.

"Where'd you come from anyway?" Anselmo asked. "Jim Player said something about flying saucers or some such. Now I don't hold much truck with them things, but y—"

Anselmo strangled on the last word. His eyes had riveted on his hand, the hand he touched the animal's head with seconds earlier.

His fingers were turning purple.

He had a fleeting desire to turn and run. It passed quickly. After a moment, the animal raised its head to look at Anselmo for the first time.

In a distinctly questioning tone, it said, "Moo?"

"Moo," Anselmo answered.

There were two purple cows grazing on the hillside when Jim Player arrived from town a few minutes later.

POSTREADING VOCABULARY EXERCISE

Directions

Complete each of the sentences below by filling in the blank with the appropriate term from the following list.

hallucination	chlorophyll	grazing
foreman	oblivious	corrugated

Jake, the _____ of the ranch, waited angrily for the wranglers to show up.

"Sorry we're late, boss," they said, as their pickup skidded to a stop. "You wouldn't believe what we saw."

Jake rolled the woodpick from one corner of his mouth to the other and pushed back his hat. "Are you

boys _____ to time, or something? We've got work to do and you are hours late."

"The road was blocked, boss," the driver spoke. "Just as we came around Shady Bend we saw this

six-foot frog _____ in the south pasture."

"We almost missed him because he was just as green as the _____ in the grass," one of the other men added. "Strangest thing I ever saw."

"You boys must be having an _____. Frogs don't graze on grass; they nab flies." Jake pulled off his glasses and began cleaning them with his shirttail. "Now get that green hay out of the pickup and spread it around the ground to dry."

A wrangler's forehead became _____ in surprise as he looked toward the truck. "That isn't hay in the back of the truck, boss."

THINKING ABOUT WHAT YOU'VE READ

1. When Floyd saw the purple cow in his field, he thought that his mind was playing tricks on him. Why did he find it so hard to believe what he had seen?
2. What was the reaction of other people when Floyd mentioned the purple cow to them?
3. Imagine and describe what the same scene at the Anselmo farm might be later that afternoon.

COMPREHENSION— referents

Directions

Jim Player never did meet Floyd on the hill . . . at least not in the form he thought he would. Jim, in preparing his story about the strange disappearance of Floyd and the even stranger appearance of two purple cows, had to make certain that he had all of the facts straight. When writing the story, he had to be careful to present the facts to his readers without confusing them. Read the following sentences and answer the questions about each.

1. When Floyd saw the purple cow on the hillside, he thought *his* mind must be hallucinating. *He* pulled his truck to the side of the road and looked again. *It* was still there.

   Who does the word *his* refer to? _____

   Who does the word *he* refer to? _____

   What does the word *it* refer to? _____

2. Anselmo walked to the telephone. Picking **It** up, he phoned Jim Player, the editor of the local newspaper. When **he** answered, Floyd told him about the purple cow.

    It refers to (check one):

    ☐ the purple cow   ☐ the newspaper   ☐ the telephone

    He refers to (check one):

    ☐ Jim Player   ☐ Floyd Anselmo   ☐ the purple cow

3. Floyd Anselmo said slowly, "I don't know anything about flying saucers. All I know is there's a purple cow on one of my hillsides. If **you** want to come out **here**, I'll show **it** to **you**." (Insert correct letter of following choices on lines below.)

    A. the cow   B. the farm   C. Jim Player   D. Floyd Anselmo

    The word **you** refers to _____

    The word **It** refers to _____

    The word **here** refers to _____

COMPREHENSION— recall of information

Directions

In any newspaper article it is not only important to get the facts straight but to get them in the right order as well. Read the following sentences. First determine if the sentence describes something that actually happened in the story. If it did **not** happen, place an **X** on the line. If it did happen, then number the sentence in the order it occurred in the story.

1. ____ One cold morning a purple cow wandered up to Floyd.
2. ____ Floyd's wife was upset when she learned that her husband was about to phone the newspaper about the cow.
3. ____ One cold morning Floyd saw a purple cow grazing on the hillside.
4. ____ Floyd's wife Amy and the foreman returned to the pasture with Floyd.
5. ____ Jim Player agreed to come out and look at the cow.
6. ____ Jim Player saw the UFO drop the cow onto the pasture.
7. ____ Floyd's wife hoped that her husband would get his picture in the paper.
8. ____ There were two cows in the pasture when Jim Player arrived.
9. ____ Floyd Anselmo returned home to tell his wife Amy and the foreman about the cow.
10. ____ Jim Player knew where he could sell the two purple cows.

ACADEMIC SKILLS— table of contents

Newspapers, like any written material, must be organized into sections that can be found easily by the reader. Stories such as the one Jim Player certainly will write about the purple cow are usually placed in the section of the paper reserved for human interest stories—articles that entertain as well as inform the reader. If everyone in town began to disappear, then the story would quickly be moved to the front page and become a major news item.

Directions

Listed below are sixteen headlines and four major sections common to many daily newspapers. Complete the outlines following the list by indicating first the major section heading and then the four stories that would appear in that section. Once you have done this, prepare a table of contents for the newspaper on a separate sheet of paper.

Sports
Moose Herd Moves into Alaska
  Hospital Parking Lot
Canning Tomatoes Can Be Tricky
New Soccer Stadium to Be Built
Business
Tips to Keep Grasshoppers
  Out of Your Garden
Rogers Pitches No-Hitter
Associated Profits Rise 1/2%
Alden Captures Marathon Title

Real Estate Sales Up
Aspen Ski Area Plans Expansion
Video Game Profits Dip
Insulate Now to Avoid Winter Chills
Today
Creature Spottted in Great Lakes
Unidentified Lights Glow Over City
AWA Air and National Lines to Merge
Confused Pilot Lands at Wrong Airport
Home and Garden

_____   _____

_____   _____

_____   _____

_____   _____

_____   _____
_____   _____
_____   _____
_____   _____
_____   _____

WRITING SKILLS—writing
a newspaper story

Journalists, those who write for newspapers and other news publications, have developed a particular style of writing. Newspaper articles are written in a style which gives the readers the greatest amount of information in the least amount of space. The first paragraph of a news story typically will describe *what* happened, *where* and *when* it occurred, *why* the incident took place, and *who* was involved. Jim Player's story about the purple cow might well begin:

County resident Floyd Anselmo vanished early Monday morning after reporting the strange presence of a purple cow in his pasture. Authorities responding to Anselmo's call found no sign of the farmer but did locate two purple cows on his land. The investigation is continuing.

Directions

Using a story from one of the newspaper sections in the previous exercise, write a complete news article, starting with the headline and your name as the reporter. Remember to present as much information as possible, in the correct sequence. Quotes from knowledgeable sources would be helpful. If needed, continue the story on a separate sheet of paper.

_____

By _____

_____

_____

_____

_____

_____

_____

_____

_____

_____

_____

_____

_____

_____

_____

_____

_____

_____

_____

JUST CALL ME IRISH

Richard Wilson

Prereading Vocabulary

tenant: one who pays rent to occupy the property of another; a resident. *(One of the tenant(s) in the new housing project was Irish.)*

spiel: a convincing speech; a sales pitch. *("Happy" Horman felt stupid giving his spiel to a setter.)*

lap: to eat or drink using only the tongue. *(The setter's son spent much of the day lap(ping) up the water from the leaking faucets.)*

hifalutin: to act as if you were better than others; condescending. *(Although he had manners, Irish did not consider himself a hifalutin dog.)*

homo sapiens: modern people. *(Irish found it easy to live among the homo sapiens.)*

caste system: a social arrangement that separates people according to their wealth or background. *(Irish felt that he did not get promoted because of the caste system that existed in the army.)*

hackles: the feathers on the neck of a bird or the hair on the back of an animal's neck. *(When Irish became angry, it raised the hackles on his neck.)*

indignity: something that offends one's pride; a humiliation. *(Irish suffered many indignit(ies) while serving in the army's K-9 Corps.)*

The housing development near the university was newly finished. Salesman John F. ("Call Me Happy") Horman had waited a week for the tenants to become settled before making the rounds with his sample electric rat trap and his order book. He began at the southwest corner of the project and knocked at the first door.

As it opened Happy went into his spiel. Toward the end of his second sentence, he skidded to a stop in the middle of a syllable when he realized he was talking to a dog. A female dog.

Happy was confused. "Is your master in?" he asked.

"Just a minute," said the dog.

The door closed and Happy stared hard at it. Then it opened again. A larger dog stood there.

"What can I do for you?" asked the larger dog.

"This is ridiculous," said Happy. "When I asked that other dog if her master was in, I meant the master of the house, not *her* master." He consulted the list of names of the families who lived in the project. "I was looking for a Mr. Setler."

"Setter's the name," said the dog. "They misspelled it. I'm the master of the house. Is there something I can do for you?"

"I don't know." Happy Horman took off his glasses, wiped them, replaced them on his nose, replaced his handkerchief and looked at the large reddish animal in the doorway. "This is very strange. Are you a talking dog?"

"Obviously." The dog swatted a fly with his tail. "Are you a talking man?"

"Why—yes."

"Then why don't you say something? Are you with the housing project? Because if you are, I wish you'd do something about the sink. It leaks. And my son Whiffet is getting tired of lapping up after it. Besides, I think it's undignified."

"Mr.—ah—Setter," said Happy, mustering his faculties, "I represent the Ohm Electric Rat Trap Company. Our slogan is 'No 'ome should be without one.'" He laughed emptily. "I think you'd be interested in a little demonstration I'd like to make for you. That is, I *think* you'd. . ."

The door opened wider, and the dog who had first spoken to Happy appeared. "Irish, dear," she said, "will you please come in or go out? The kennel's getting cold."

"House, Maureen, not kennel."

"House, then. But why not ask the gentleman in?"

"Yes, won't you come in, sir?" said Irish. "If you don't mind the place being somewhat littered."

Happy went in and sat on the edge of a normal wooden chair. He looked around with interest but so far as he could see the furnishings were those of an average dwelling. It did not look at all like a doghouse, though it unquestionably was a *dog's* house.

Irish curled himself comfortably on a couch while Maureen excused herself, saying it was time the younger whelps were fed. "I'll be glad when they're weaned," she said. Happy Horman blushed.

"Mr. Setter," Happy said, "please forgive me if I seem curious, but just how—that is, why, uh—how come you're living here?"

"Why not?" Irish said. "I'm eligible."

"But I thought these houses were set aside for veterans?"

"I'm a veteran," the dog said. "Want to see my honorable discharge from the K-9 Corps?"

"Oh. But you have to be a student, and you have to be married, I thought."

"I *am* married, sir," Irish said in a hard voice. "You don't think I'm just living with her, do you?"

Happy coughed in embarrassment.

"Please, Mr. Setter, I meant to imply no such thing. But how can you be a student? At the university, that is? I realize that we're all students of human nature, heh heh, you especially, of course, being a—a canine."

"Dog is good enough. No need to get hifalutin. Would you like to hear the whole story?"

"Why, yes, I would."

"It began about 1949," Irish said, settling himself more comfortably. "My master (before I became my own) was Professor Neil Wendt, the big nuclear physics man on campus. Or should I call him the nuclear physics homo sapiens?" he asked archly.

Happy laughed hollowly.

"I don't fully understand, even now, what exactly Wendt was doing but I was his constant companion, his dumb animal friend. Then one day, as I reconstruct it, I was affected by radiation and when Wendt called me I said 'Coming.' Just like that. I don't know who was more surprised, Wendt or me.

"After some preliminary confusion we sat down and talked the thing out. We found that we could be of considerable help to each other. I suggested a few improvements in his equipment, having had a dog's eye view of it from underneath; though actually it made little difference because in a few weeks the Atomic Energy Commission took the whole thing over. In the meantime he went with me to the dean and with a little coaching I was able to pass the examinations and was awarded a bachelor's degree. You a college man, sir?"

"Er, no," Happy said.

"Um. Well, later, when I was working toward my master's I realized there were more important things than books. I refer to the Korean War. So, as any red-blooded American dog would do, I enlisted. The K-9 Corps is a fine organization, in its limited way, and I was very quickly promoted to sergeant. But the caste system! Absolutely unfair. I had heard about openings in Officer Candidate School and inquired about them. My first sergeant laughed at me but by dogged persistence I got to see the regimental commander.

"He was sympathetic but had to refuse my application. Said there was nothing in the ARs about it. What a welter of dogma those army regulations are! So I was forced to finish out my army career as an enlisted dog. True, I finally made master sergeant—though they claimed it was stretching a point for a dog to become a master—but my hackles still rise when I think of the indignities I suffered under the myth of racial superiority. What a blow to one's pride to be forced to write 'animal' opposite the word race, when almost everyone else was able to write 'human.'"

Irish glared so at Happy that the salesman winced.

"But that's all over now, Mr. Setter," Happy said. "And now you're back at school. What field are you in?"

"Anthropology, of course," Irish said. "But we've talked enough about me. What was it you had to show me, sir?"

"I really don't think you'd be interested," Happy said. "It's something *you* certainly would have no use for. You see, it's a rat catcher, and surely you of all ani—er, of all people, wouldn't—"

"Well, I don't know. I don't see why not. I suppose you might argue that I'm perfectly capable of catching rats myself. It's true that I'm still a young dog, but I don't have the time for sport that I used to. Suppose I take a look at your model."

Relieved to be in action again, the salesman rose and plugged in the cord of his electric rat trap. With a rubber rat he demonstrated its possibilities.

"Well, I'll be doggoned," Irish said. "Maureen, come in and look at this gadget."

The female dog (as Happy preferred to think of her) came in. She also marvelled at its efficiency. "Let's get one, Irish," she said. "It'll save us an awful lot of work."

"I think I will," Irish said. "If you'll make out an order for us, sir? That's fine. Just put the pen in my teeth and I'll sign it. There."

Happy handed over the receipt, discreetly wiped the doggy saliva from his pen and prepared to go.

"Drop in any time," Irish said. "You might like to come in some evening and tear a bone with us."

Happy forced a chuckle. "You're quite a wag, Mr. Setter," he said daringly, and was relieved when his customer broke into a barking laugh and closed the door after him.

Happy Horman took several deep breaths of air, then turned back to look at the house. No one was visible behind its windows. He looked at his order book. There was the bold signature: *I. Setter.* He shook his head, shrugged, and went to the next house.

He knocked. A fat young man opened the door.

"I beg your pardon," Happy said, "but is your dog in?"

POSTREADING VOCABULARY EXERCISE

Directions

Fill in the circle beside the one word that does not fit with the rest. Refer to your dictionary if necessary.

1. *indignity*
   ○ embarrassment
   ○ humiliation
   ○ humorous
   ○ abuse

2. *tenant*
   ○ renter
   ○ occupant
   ○ inhabitant
   ○ owner

3. *hackles*
   ○ hair
   ○ feathers
   ○ scales

4. *spiel*
   ○ recording
   ○ speech
   ○ lecture
   ○ talk

5. *hifalutin*
   ○ pompous
   ○ shy
   ○ condescending
   ○ boastful

6. *lap*
   ○ drink
   ○ slurp
   ○ eat
   ○ sit

7. *homo sapiens*
   ○ man
   ○ canine
   ○ human being
   ○ person

THINKING ABOUT WHAT YOU'VE READ

1. What were Irish's feelings about having to live among human beings?
2. As hard as he tried, Irish was unable to get promoted in the army. What did he believe were the army's reasons for not making him an officer? What do **you** believe its reasons were?
3. Happy Horman made a sale to the Setter family, and Irish complimented the man on his sales ability. What are some of the skills necessary to become a successful salesperson?

COMPREHENSION— recognizing character emotion

Directions

Dogs are very perceptive animals. They are capable of sensing not only fear and danger, but also the emotions of human beings. It didn't take Irish long to realize just how nervous Happy Horman was to be speaking with a dog. Read the following selections from the story. Complete each sentence by checking the emotion that corresponds with the selection.

1. Horman took off his glasses and looked at the large reddish animal and asked, "Are you a talking dog?"
   We can conclude that Horman was:
   ☐ happy   ☐ surprised ✓   ☐ bored

2. "Obviously," the dog replied, swatting a fly with his tail. "Are you a talking man?"
   Irish had been asked that question so many times that he had grown _____ with it.
   ☐ bored ✓   ☐ angry   ☐ humiliated

3. Irish reported that the sink in his new apartment leaked and that his son Whiffet was getting tired of lapping up the water. "Besides," Irish continued, "I think it is undignified to have to lap up spilled water."
   How do you think Irish felt about his new apartment?
   ☐ irritated ✓   ☐ pleased   ☐ enthusiastic

4. Irish enlisted in the United States Army's K-9 Corps. He was quickly promoted to sergeant. He was not, however, allowed to enroll in Officer Candidate School because no animal had ever been enrolled in that school.
   What emotion did Irish probably feel after being turned down?
   ☐ discriminated against ✓   ☐ proud   ☐ peaceful

5. Irish and his wife Maureen looked at Happy Horman's rat catcher. After several minutes, they decided to buy one. How do you think Happy felt when Irish asked for a pen to be put in his mouth?
   ☐ surprised ✓   ☐ disappointed   ☐ angry

COMPREHENSION—analyzing dialogue

Directions

John F. ("Call Me Happy") Horman earned his living by convincing people to buy his goods. To be successful in this field, Horman had to be a persuasive speaker as well as a good listener. Speaking and listening are two skills used in conversation. Read the dialogue below from the story and answer the questions that follow.

The housing development near the university was newly finished. Salesman John F. ("Call Me Happy") Horman had waited a week for the tenants to become settled before making the rounds with his sample electric rat trap and his order book. He began at the southwest corner of the project and knocked on the first door.

As it opened, Happy went into his spiel. Toward the end of his second sentence, he skidded to a stop in the middle of a syllable when he realized he was talking to a dog. A female dog.

Happy was confused. "Is your master in?" he asked.

"Just a minute," said the dog.

The door closed and Happy stared hard at it. Then it opened again and a larger dog stood there.

"What can I do for you?" asked the larger dog.

"This is ridiculous," said Happy. "When I asked the other dog if her master was in, I meant the master of the house, not her master." He consulted the list of names of families who lived in the project. "I was looking for a Mr. Setler."

"Setter's the name," said the dog. "They misspelled it. I'm the master of the house. Is there something I can do for you?"

"I don't know." Happy Horman took off his glasses, wiped them, replaced them on his nose, replaced his handkerchief, and looked at the large reddish animal in the doorway. "This is very strange. Are you a talking dog?"

"Obviously." The dog swatted a fly with his tail. "Are you a talking man?"

1. What does the expression "skidded to a stop" mean?

_____

_____

2. Happy Horman demonstrated some excellent points of salesmanship. Circle the one item that does **not** support positive salesmanship.

   A. Happy waited a week for the tenants to get settled before he began his sales calls.

   B. Happy had his order book ready.

   C. Happy said it was ridiculous to be talking to a dog.

   D. Happy brought a demonstration model of his product with him.

3. How do you think Happy felt when a dog answered the door?

_____

_____

4. How do you think Happy felt when Irish opened the door?

_____

_____

5. When we speak with others, **how** we say something is often as meaningful as what we say. Rewrite Happy's final sentence in a more positive fashion.

_____

_____

6. Dialogue and action are often combined in the same sentence, giving the reading greater meaning What did Happy do when he first began speaking with Irish?

_____

_____

7. What was Irish doing as he asked Happy if he was a talking man?

_____

_____

LIVING SKILLS— completing an order form

We depend on each other to supply the goods and services we cannot make for ourselves. Before the development of money, people bartered, or traded with their neighbors, for the things they needed. As societies became more complex, people in the community specialized in the production of certain products which they then sold in markets as they traveled from town to town.

Today, most sales are conducted in stores and money is used to pay for the goods. Some sort of record of the sale is usually kept by the merchant. The person buying the product is given a copy or receipt of the sales record. If the sale involves a product that is to be delivered later, an order form is completed. Order forms indicate the color, size, quantity, and cost of the item to be delivered. Designer rat traps are no exception.

Directions

You have become the Senior Sales Trainer for
Dealco Designer Rat Traps, Ltd. Your job is to demonstrate to your student
sales staff how to sell your product. You have just sold a rat trap on your first stop.
Read the dialogue below and then fill in the order form that follows.

"Good morning, sir. I represent Dealco Designer Rat Traps, Ltd."

"I thought you'd never get here," your customer replies. "I need eight rat traps as soon as possible.
I'll take four of your deluxe traps and four of your basic models in black."

"Certainly, certainly, Mr. ....."

"Eldridge. Seymour Eldridge," your customer replies. "Now, how much do I owe you? I'll pay you cash.
I'll pay half now and the rest when I take delivery."

"The deluxe traps are $6.50 each. The basic traps are on special this month. They have a unit price
of $3.00 apiece."

"Fine," Eldridge replies. "Send them to me at my office at 860 West Moorhead. That's across
the state line in Boulder, Colorado 80302. If you have any problems, phone me at (303) 555-2051."

ORDER #

Date of Order: _____

Customer Name _____
| | LAST | FIRST | MI |

Home Address | | | | | | | | | | | | | | | | | | | | | | | | | | | | | | | | | | |

| | | | | | | | | | | | | | | | | | | | | | | | | | | | | | | | | | | |
| | CITY | STATE | ZIP |

Number of Traps Ordered	Style	Unit Price	Total Price

Method of Payment
☐ Cash   ☐ Check

Delivery Date: _____

Total	
Deposit	
Balance	

WRITING SKILLS— writing dialogue

Your last sale was simple. The next one, however, will not be quite so easy.

Directions

Pick a partner for this activity. You must communicate only through written discussion, and you must convince your partner to buy a rat trap. Begin by introducing yourself and then start your sales spiel. For each statement you make, your partner must respond. The beginning dialogue has been done for you.

Salesperson: "Good morning. I represent the Dealco Designer Rat Trap Company. I would like to demonstrate our latest model."

Customer: "_____

_____"

Salesperson: "_____

_____"

Customer: "_____

_____"

Salesperson: "_____

_____"

(Continue on a separate sheet of paper.)

ZOO
Edward D. Hoch

Prereading Vocabulary

professor: a teacher or instructor. *(Professor Hugo was the owner of the Interplanetary Zoo.)*

wonderment: surprise, astonishment. *(The visitors to the zoo stared at the creatures with* wonderment.*)*

breed: a kind or group of like beings. *(The Earthlings had never seen such a strange* breed *of alien before.)*

chatter: to speak rapidly. *(The zoo creatures seemed to engage in nonstop* chatter.*)*

horrified: a feeling of intense fear; frightened. *(The Earthlings were both fascinated and* horrified *by the small spider creatures.)*

fascinated: held an interest in; spellbound. *(Professor Hugo's zoo* fascinated *everyone who visited it.)*

jagged: an uneven surface; rough. *(The spider creatures lived among the* jagged *rocks of their planet.)*

offspring: the children of a particular parent. *(The spider creature was happy to see the return of her* offspring.*)*

The children were always good during the month of August, especially when it began to get near the twenty-third. It was on this day that the great silver spaceship carrying Professor Hugo's Interplanetary Zoo settled down for its annual six-hour visit to the Chicago area.

Before daybreak the crowds would form, long lines of children and adults both, each one clutching his or her dollar, and waiting with wonderment to see what race of strange creatures the Professor had brought this year.

In the past they had sometimes been treated to three-legged creatures from Venus, or tall, thin men from Mars, or even snake-like horrors from somewhere more distant. This year, as the great round ship settled slowly to earth in the huge tri-city parking area just outside of Chicago, they watched with awe as the sides slowly slid up to reveal the familiar barred cages. In them were some wild breed of nightmare — small, horse-like animals that moved with quick, jerking motions and constantly chattered in a high-pitched tongue. The citizens of Earth clustered around as Professor Hugo's crew quickly collected the waiting dollars, and soon the good Professor himself made an appearance, wearing his many-colored rainbow cape and top hat. "Peoples of Earth," he called into his microphone.

The crowd's noise died down and he continued. "Peoples of Earth, this year you see a real treat for your single dollar — the little-known horse-spider people of Kaan — brought to you across a million miles of space at great expense. Gather around, see them, study them, listen to them, tell your friends about them. But hurry! My ship can remain here only six hours!"

And the crowds slowly filed by, at once horrified and fascinated by these strange creatures that looked like horses but ran up the walls of their cages like spiders. "This is certainly worth a dollar," one man remarked, hurrying away. "I'm going home to get the wife."

All day long it went like that, until ten thousand people had filed by the barred cages set into the side of the spaceship. Then, as the six-hour limit ran out, Professor Hugo once more took microphone in hand. "We must go now, but we will return next year on this date. And if you enjoyed our zoo this year, phone your friends in other cities about it. We will land in New York tomorrow, and next week on to London, Paris, Rome, Hong Kong, and Tokyo. Then on to other worlds!"

He waved farewell to them, and as the ship rose from the ground the Earth peoples agreed that this had been the very best Zoo yet...

Some two months and three planets later, the silver ship of Professor Hugo settled at last onto the familiar jagged rocks of Kaan, and the queer horse-spider creatures filed quickly out of their cages. Professor Hugo was there to say a few parting words, and then they scurried away in a hundred different directions, seeking their homes among the rocks.

In one, the she-creature was happy to see the return of her mate and offspring. She babbled a greeting in the strange tongue and hurried to embrace them. "It was a long time you were gone. Was it good?"

And the he-creature nodded. "The little one enjoyed it especially. We visited eight worlds and saw many things."

The little one ran up the wall of the cave. "On the place called Earth it was the best. The creatures there wear garments over their skins, and they walk on two legs."

"But isn't it dangerous?" asked the she-creature.

"No," her mate answered. "There are bars to protect us from them. We remain right in the ship. Next time you must come with us. It is well worth the nineteen commocs it costs."

And the little one nodded. "It was the very best Zoo ever. ..."

POSTREADING VOCABULARY EXERCISE

Directions

Read the following questions and statements and put the correct answer in the proper spaces below. Refer to the Prereading Vocabulary for your answers.

1. What is another name for an instructor?
2. What word means something very different from silence?
3. Interested and _____ mean about the same thing.
4. A person who is surprised by an experience is filled with (choose one): wonderment   suspicion
5. You are your parents' _____.
6. Clydesdales are a large _____ of horse.
7. The _____ edge of the glass cut his hand.
8. Maxine was _____ when she saw the Halloween mask.
9. Where is the largest zoo in the United States?

THINKING ABOUT
WHAT YOU'VE READ

1. Professor Hugo made a great deal of money in his business. Not only did he charge the visitors to his intergalactic zoo, he also charged the aliens he carried on his ship. How was he able to charge both groups?

2. Given the choice, would you rather have been on board the Professor's ship visiting several galaxies, or on Earth observing strange aliens inside the space vessel?

3. Do you feel there was anything wrong with the way Professor Hugo conducted his business? Why?

COMPREHENSION— stating correct sequence

Directions

Professor Hugo had to develop a very efficient schedule if his interplanetary zoo was to remain successful. Read the following sentences and arrange them in the order in which they occurred in the story.

1. _____   August 23   The sides of the large ship slide up to reveal the familiar cages.
2. _____   August   Professor Hugo picks up a group of tourists on the planet Kaan.
3. _____   August 23   People begin lining up in the tri-city parking area.
4. _____   October   Professor Hugo lands once again on Kaan.
5. _____   August 23   Professor Hugo's ship lands outside Chicago.
6. _____   August   Professor Hugo's ship lands in New York.
7. _____   August 23   Professor Hugo addresses the people of Earth.

COMPREHENSION— cause and effect

Directions

Many people are strong believers in the old saying, "Nothing succeeds like success." Professor Hugo was indeed a successful man. Not only did he run an intergalactic tour service, he also operated a very popular zoo. To become successful, Professor Hugo must have realized that everything he did or *caused* to happen had some effect on his business. Using the story as your guide, read each of the causes identified below and match it with the proper effect.

CAUSE	EFFECT

CAUSE

1. _____ Because Professor Hugo's zoo came at the end of every August, . . .

2. _____ Because Professor Hugo's zoo could remain in Chicago for just six hours, . . .

3. _____ Because Professor Hugo was about to address the crowd, . . .

4. _____ Because they were surprised to see the Earthlings crowded around the ship, . . .

5. _____ Because the creatures aboard Professor Hugo's ship were so unusual, . . .

6. _____ Because the she-creature was excited to see the return of her offspring, . . .

EFFECT

A. the spider-like creaturezs chattered constantly.

B. the crowd grew immediately quiet.

C. she babbled greetings in a strange tongue.

D. the children always behaved themselves during the month.

E. people lined up before dawn.

F. the Earthlings reacted in horror and wonderment.

ORGANIZATION SKILLS— planning a schedule

The success of an operation such as Professor Hugo's depends on a carefully planned schedule. The effects of arriving too early in a city could be just as disastrous as arriving too late. Any business that hopes to succeed at being in several places at different times must develop a flawless schedule, as you will soon discover. You have been appointed to the position of Promotions Director for Cosmos to Cosmos Traveling Circus. Your first job is to develop the travel schedule for the circus.

Directions

Using the information below, develop a complete schedule. Record the information in the chart that follows.

"Our tour begins in June this year," Doctor Presente, the circus owner, began. "We will be including a herd of trained zebrats as a special attraction this year. They are being loaned to us by the Royal Martian Society for Veterinary Science on the condition that *we* deliver them to the Saturn Regional Zoo in August. We must arrive on Mars on June 1."

Randy Smsa, the traffic coordinator, nodded. "That will work. If we leave Mars on June 2, we can arrive in the matched orbit of Earth's Lunar Colony Number Six on June 15."

"Our first performance there is on June 16," the ringmaster added. "It is a benefit for the Save the Wildebeest Fund."

"That's news to me," Rhonda Billet said. "I should be informed of such things. Benefit tickets sell for $25.00 a seat while regular performances are booked at $10.00 a seat. I'll have to have the tickets reprinted."

"It's an added expense, but worth our time," the doctor said. "Good public relations are important. Is our next stop Earth?"

"Negative," Smsa continued. "We arrive on Earth on July 1 and July 22. When we leave the Lunar Colony on June 18, we fly straight to Venus for a performance on June 20. We'll arrive around 10 p.m. the night before. We'll move on to Earth on June 28."

"I would like to schedule a benefit on Uranus," the doctor said. "Can we be there on July 17?"

"Absolutely not!" Camden, the chief mechanic, answered. "The ships must arrive at the Neptune repair docks on July 17. That's the only date I could get. We must leave Earth on July 15."

"And again on July 28 if we are going to deliver the zebrats on August 1," the doctor concluded. "Let's put the schedule together."

SCHEDULE OF COSMOS TO COSMOS TRAVELING CIRCUS			
Date of Arrival	Date of Departure	Planet	Reason for Stop

WRITING SKILLS— planning a promotional campaign

Businesses that travel from one location to another do not have the luxury of building a permanent home in each city where customers can come. Therefore, it is necessary for traveling businesses such as the Cosmos to Cosmos Circus to advertise when they are coming to town. The Promotional Director has the task of following the identical schedule as the circus, but traveling a few days ahead of the group. On each planet you must decide the best way to promote, or advertise, the circus. You can spend your advertising budget of $1,000.00 on three types of advertising—television, radio, or print ads. Television advertising costs $24.00 a second, radio costs $15.00 a second, and print ads sell for $750.00 a page. Spend your money in any combination you like.

Directions

Pick one stop on the schedule in the previous exercise and plan the advertising budget below. You must spend within $20.00 of your budget. When you have planned your budget, write and design the advertisements on separate paper.

ADVERTISING BUDGET

Planet _____

Beginning Balance .... $_____

Advertising Expenditure for _____

Amount .... $ _____

Advertising Expenditure for _____

Amount .... $ _____

Advertising Expenditure for _____

Amount .... $ _____

Ending Balance .... $ _____

Directions

Since you are paying for your radio and television advertising by the second, the lines below are divided into five-second segments. You will have to practice your lines to determine whether you can fit your advertisement into the time you bought.

0:00 _____

_____

0:05 _____

_____

0:10 _____

_____

0:15 _____

_____

0:20 _____

_____

0:25 _____

_____

0:30 _____

_____

(Continue on a separate sheet of paper.)

DOG STAR
Mack Reynolds

Prereading Vocabulary

abundance: plenty. *(The planet Sirius had an abundance of pitchblende.)*

delegate: a person sent to act as a representative for others. *(The rulers of Sirius sent their* delegate *to meet with the visitors from Earth.)*

tentacles: long, flexible extensions of a body used to grasp and feel objects. *(Since Sirians had no arms, they used the* tentacles *of an octopus to move and grasp objects within their reach.)*

indignant: angry as a result of an injustice. *(The leader of the Earth visitors was* indignant *when he learned that the Sirians thought dogs were more intelligent than people.)*

domesticated: tame, housebroken. *(The Sirians thought that the men from Earth were the* domesticated *pets of the dog, Gimmick.)*

comprehend: to understand. *(It was difficult for Lt. Grant to* comprehend *that the inhabitants of Sirius looked just like the dogs on Earth.)*

When man's first representatives landed on Sirius Two, hungry to trade for that planet's abundance of pitchblende, they carried with them, as ship's mascot, one of the few dogs left on Earth.

That Gimmick was one of the very last was not due to disease, nor reproductive failure. It was just that man was going through a period of wearying of his ages-long companion. The Venusian *marmoset*, the Martian *trillie*, were much cuter, you know, and much less trouble.

Captain Hanford — leader of the three-man, one-dog crew — saluted the Sirian delegates snappily, only mildly surprised at the other's appearance. In a small saddle, topping what appeared almost identical to an Earthside airedale, was an octopus-looking creature. It was not until later that the captain and his men realized that the dog-like creature was the more intelligent of the two, and the octopus was a telepathically controlled set of useful tentacles.

Telepathic communication can be confusing, since it is almost impossible, when the group consists of several individuals, to know who is "talking."

When the amenities had been dispensed with, the Sirian leader remarked in friendly fashion, "I would say our domesticated animals were somewhat superior to yours. Eight tentacles would seem more efficient than two five-fingered limbs, such as yours possess."

Captain Hanford blinked.

"In fact," the Sirian continued, somewhat apologetically, "if you don't mind my saying so, your creatures are somewhat repulsive in appearance."

"I suppose we are used to them," Hanford replied, swallowing quickly.

Later, in the space ship, the ship's captain looked at his men indignantly. "Do you realize," he said, "that they think Gimmick is the leader of this expedition and that we're domesticated animals?"

Ensign Jones said happily, "Possibly you're right, Skipper, but they were certainly friendly enough. And they sure came through nicely on the uranium exchange deal. The government will be pleased as..."

Hanford insisted, "But do you realize what those Sirians would think if it came out that Gimmick was a pet? That we consider him an inferior life form?"

Lieutenant Grant was the first to comprehend. "It means," he said slowly, "that from now on, every time we come in contact with Sirians, a dog is going to have to be along. It means that every ship that comes for a load of pitchblende, is going to have to have several. We've got to continue pretending that the dog is Earth's dominant life form, and man his servant. Every time we *talk* to a Sirian, we're going to have to pretend it's the dog *talking*."

"Holy Smokes," Jones said, "there aren't that many dogs left on Earth. We're going to have to start breeding them back as fast as we can."

The Captain looked down to where Gimmick, his red tongue out as he panted so that he looked as though he was grinning, lay on the floor.

"You son-of-a—" the Captain snapped at him.

But Gimmick's tail went left, right, left, right.

POSTREADING VOCABULARY EXERCISE

Directions

Many of the words we use in our English language came originally from ancient Latin or Greek. Match each of the vocabulary words in the first column with the letter of its Latin or Greek form (in the second column.) Refer to your dictionary for further help.

____ 1. domesticated

____ 2. abundance

____ 3. indignant

____ 4. tentacles

____ 5. delegate

A. *indignari*: to regard as unworthy; *in-*, not + *dignus*, worthy (Latin)

B. *domesticus* < *domus*: house (Latin)

C. *delegatus* < *delegare*: to dispatch; *de-*, away + *legare*, to send < *lex*, law (Latin)

D. *abundare*: to overflow (Latin)

E. *tentaculum* < *tentare*: to touch (Latin)

THINKING ABOUT WHAT YOU'VE READ

1. Now that the people of Earth and the planet Sirius have come to an agreement and contact between the two will increase, what will happen to the population of dogs on Earth?
2. The Sirians believed that Gimmick was the leader of the expedition from Earth. What reasons did they have for believing as they did?
3. What purpose did the Sirians assume the men from Earth played?

COMPREHENSION— distinguishing between true and false statements

Directions

Although Captain Hanford deceived his hosts by allowing them to assume Gimmick was in command, the deception would not have lasted. The Sirians conducted a full investigation into their new business partners. Some of the information collected was true, while the rest was false. Read the following statements. If the statement is true, place a *T* in front of the sentence. If it is false, place an *F* before the statement.

_____ 1. At the time of the Sirius-Earth pitchblende deal, dogs were very popular on Earth.
_____ 2. Captain Hanford and his crew planned to capture the Sirians and use them as pets on Earth.
_____ 3. Gimmick was able to control the movements of the humans by using telepathy.
_____ 4. Captain Hanford was the leader of the Earth delegation.
_____ 5. Venusian marmosets had become more popular on Earth than dogs.
_____ 6. The Earth delegation initially thought that the Sirian delegate was the octopus.
_____ 7. The Earth delegation, realizing the importance Gimmick played in the pitchblende deal, planned to bring a dog with them on future visits.

___ 8. Captain Hanford was disturbed by the fact that the Sirians could have thought a mere dog was more intelligent than he.

___ 9. If the Earth delegation's deception was to work, they would have to start breeding dogs as soon as possible.

___ 10. Despite the confusion, the delegations did sign an agreement to trade pitchblende for uranium.

COMPREHENSION— forming assumptions

Since Gimmick looked like them, the Sirians assumed he was the leader of the delegation from Earth. And since they communicated telepathically, the Sirians assumed it was the dog that was speaking. Imagine their surprise the first time Gimmick barked, bit, or buried a bone. They would then discover their original assumption had been incorrect.

Directions

Below are four assumptions drawn from the story.
Read each assumption, then choose the two reasons why the assumption could be formed. Place an **X** beside your choices.

1. At first, Captain Hanford assumed that the octopus was the Sirian leader because:
   A. ___ the octopus rode in a saddle.
   B. ___ the octopus spoke several languages.
   C. ___ the octopus moved its tentacles as if it were in command.
   D. ___ the Earthmen had seen pictures of the Sirians.

2. The Sirians assumed that the men were inferior domesticated animals because:
   A. ___ the men wore strange clothing.
   B. ___ the men had just two hands with five digits each.
   C. ___ the men were aggressive and angry.
   D. ___ the men were rather repulsive in appearance.

3. Hanford and his crew correctly assumed that the Sirians thought Gimmick was in command because:
   A. ___ the Sirians spoke about the humans as if they were stupid.
   B. ___ the Sirians spoke to Gimmick in a friendly fashion.
   C. ___ the Sirians had pets that looked like humans.
   D. ___ the Sirians had visited Earth many times before.

4. You, the reader, can assume that there weren't many dogs left on Earth because:
   A. _____ a disease killed most of the dogs on Earth.
   B. _____ man had grown tired of dogs.
   C. _____ all of the dogs had traveled to Sirius years earlier.
   D. _____ the Venusian marmoset was less trouble to man than was a dog.

ACADEMIC SKILLS— categorizing

Captain Hanford and his crew were lucky to have had Gimmick with them on the expedition to Sirius. Without their loyal mascot, the pitchblende deal might never have been signed. It seems that dogs are always around when people need them. In fact, it is hard to imagine a world without dogs. According to the American Kennel Club, there are more than 25 million dogs in the United States alone.

Many dogs are trained to perform certain jobs while others are raised to be loving family pets. The Kennel Club recognizes 121 different breeds of dogs and separates them into six categories: sporting dogs, hounds, working dogs, terriers, toy dogs, and non-sporting dogs.

Directions

Below are drawings of 12 dogs. Using an encyclopedia, place each of these dogs in their proper group following the drawings.

Basset Hound

Airedale

Cocker Spaniel

Pekingese

Boxer

Beagle

Collie

Irish Setter

Miniature Schnauzer   Pomeranian   Bulldog   Standard Poodle

SPORTING DOGS	HOUNDS	WORKING DOGS
1. _____	1. _____	1. _____
2. _____	2. _____	2. _____
TERRIERS	TOY DOGS	NON-SPORTING DOGS
1. _____	1. _____	1. _____
2. _____	2. _____	2. _____

WRITING SKILLS— gathering information

The American Kennel Club did not just throw the different breeds of dogs into six different categories. These dogs were classified according to specific strengths and traits that they share with other breeds in their group.

Directions

Using an encyclopedia or books about dogs, find out as much as you can about two breeds of dogs (do not pick two from the same group). Then, using the guide below, record your findings in notation form. The outline following the guide will help you summarize your findings.

Breed: _____

Group: _____

Hair type: ☐ silky   ☐ coarse

Approximate height: _____

Approximate weight: _____

Coloring _____

Special Abilities:

_____

_____

_____

Uses:

_____

_____

_____

_____

Breed: _____

Group: _____

Hair type: ☐ silky   ☐ coarse

Approximate height: _____

Approximate weight: _____

Coloring _____

Special Abilities:

_____

_____

_____

Uses:

_____

_____

_____

_____

Directions

Using the information you collected above,
write a paragraph summarizing your finds. Compare the appearance of the two
dogs as well as describing any differences in the way they are raised. Use additional
paper if needed.

_____

_____

_____

_____

_____

_____

_____

CREATURE OF THE SNOWS
William Sambrot

Prereading Vocabulary

grandeur: grand, magnificent. *(The grandeur of the towering mountains was a surprise to Ed McKale.)*

pinnacle: a tall, pointed formation, such as a mountain peak or building top. *(Ed realized that it would be impossible for him to reach the pinnacle of the mountain.)*

vista: a distant view as seen through an opening, such as the view between two mountains. *(Ed was careful to capture enough light when photographing the vista.)*

abyss: a deep chasm; a seemingly bottomless pit. *(The photographer became uneasy while looking into the abyss.)*

crevass: a deep crack, as in a glacier. *(Ed nearly fell into a crevass while making his way across the glacier.)*

plateau: a level area of land. *(Reaching the plateau, Ed spotted the creatures.)*

massif: a large mountain or group of connecting mountains. *(Ed was not interested in the entire Himalayan mountain range. He only wanted to explore the Gauri Sankar massif.)*

acclimatized: grew accustomed to a new environment. *(It took time for Ed to become acclimatized to the high altitude.)*

escarpment: a steep slope separating two flat surfaces. *(Ed had difficulty breathing as he moved up the escarpment.)*

Ed McKale straightened up under his load of cameras and equipment, squinting against the blasting wind, peering, staring, sweeping the jagged, unending expanse of snow and wind-scoured rock. Looking, searching, as he'd been doing now for two months, cameras at the ready.

Nothing. Nothing but the towering Himalayas, thrusting miles high on all sides, stretching in awesome grandeur from horizon to horizon, each pinnacle tipped with immense banners of snow plumes, streaming out in the wind, vivid against the darkly blue sky. The vista was one of surpassing beauty; viewing it, Ed automatically thought of light settings, focal length, color filters—then just as automatically rejected the thought. He was here, on top of the world, to photograph something infinitely more newsworthy—if only he could find it.

The expedition paused, strung out along a ridge of blue snow, with shadows falling away to the right and left into terrifying abysses, and Ed sucked for air. Twenty thousand feet is really quite high, although many of the peaks beyond rose nearly ten thousand feet above him.

Up ahead, the Sherpa porters—each a marvelous shot, gap-toothed, ebullient grins, seamed faces, leathery brown—bowed under stupendous loads for this altitude, leaning on their coolie crutches, waiting for Dr. Schenk to make up his mind. Schenk, the expedition leader, was arguing with the guides again, his breath spurting little puffs of vapor, waving his arms, pointing—down.

Obviously Schenk was calling it quits. He was within his rights, Ed knew; two months was all Schenk had contracted for. Two months of probing snow and ice; scrambling over crevasses, up rotten rock cliffs, wind-ravaged, bleak, stretching endlessly toward Tibet and the never-never lands beyond. Two months of searching for footprints where none should be. Searching for odors, for droppings, anything to disclose the presence of creatures other than themselves. Without success.

Two months of nothing. Big, fat nothing.

The expedition was a bust. The goofiest assignment of this or any other century, as Ed felt it would be from the moment he'd sat across a desk from the big boss in the picture magazine's New York office, two months ago, looking at the blurred photograph, while the boss filled him in on the weird details.

The photograph, his boss had told him gravely, had been taken in the Himalayan mountains, at an altitude of twenty-one thousand feet, by a man soaring overhead in a motorless glider.

"A glider," Ed had said noncommittally, staring at the fuzzy, enlarged snapshot of a great expanse of snow and rocky ledges, full of harsh light and shadows, a sort of roughly bowl-shaped plateau, apparently, and in the middle of it a group of indistinct figures, tiny, lost against the immensity of great ice pinnacles. Ed looked closer. Were the figures people? If so—what had happened to their clothes?

"A glider," his boss reiterated firmly. The glider pilot, the boss said, was maneuvering in an updraft, attempting to do the incredible—soar over Mount Everest in a homemade glider. The wide-winged glider had been unable to achieve the flight over Everest, but, flitting silently about seeking updrafts, it cleared a jagged pinnacle and there, less than a thousand feet below, the pilot saw movement where none should have been. And dropping lower, startled, he'd seen, the boss said dryly, "creatures—creatures that looked exactly like a group of naked men and women and kids, playing in the snow, at an altitude of twenty thousand five

hundred feet." He'd had the presence of mind to take a few hasty snapshots before the group disappeared. Only one of the pictures had developed.

Looking at the snapshot with professional scorn, Ed had said, "These things are indistinct. I think he's selling you a bill of goods."

"No," the boss said, "we checked on the guy. He really did make the glider flight. We've had experts go over that blowup. The pictures's genuine. Those are naked biped, erect-walking creatures." He flipped the picture irritably. "I can't publish this thing; I want close-ups, action shots, the sort of thing our subscribers have come to expect of us."

He'd lighted a cigar slowly. "Bring me back some pictures I can publish, Ed, and you can write your own ticket."

"You're asking me to climb Mount Everest," Ed said, carefully keeping the sarcasm out of his voice. "To search for this plateau here," he tapped the shoddy photograph, "and take pix of—what are they, biped, erect-walking creatures, you say?"

The boss cleared his throat. "Not Mount Everest, Ed. It's Gauri Sankar, one of the peaks near Mount Everest. Roughly, it's only about twenty-three thousand feet or so high."

"That's pretty rough," Ed said.

The boss looked pained. "Actually, it's not Gauri Sankar either. Just one of the lesser peaks of the Gauri Sankar massif. Well under twenty-three thousand. Certainly nothing to bother a hot-shot exparatrooper like you, Ed."

Ed winced, and the boss continued. "This guy—this glider pilot—wasn't able to pin-point the spot, but he did come up with a pretty fair map of the terrain—for a pretty fair price. We've checked it out with the American Alpine Club; it conforms well with their own charts of the general area. Several expeditions have been in the vicinity, but not this exact spot, they tell me. It's not a piece of cake by any means, but it's far from being another Annapurna, or K_2, for accessibility."

He sucked at his cigar thoughtfully. "The Alpine Club says we've got only about two months of good weather before the inevitable monsoons hit that area—so time, as they say, is of the essence, Ed. But two months for this kind of thing ought to be plenty. Everything will be first class; we're even including these new gas guns that shoot hypodermic needles, or something similar. We'll fly the essentials into Katmandu and air-drop everything possible along the route up to your base at"—he squinted at a map—"Namche Bazar. A Sherpa village which is twelve thousand feet high."

He smiled amiably at Ed. "That's a couple of weeks' march up from the nearest railroad, and ought to get you acclimatized nicely. Plenty of experienced porters at Namche, all Sherpas. We've lined up a couple of expert mountain climbers with Himalayan background. And the expedition leader will be Dr. Schenk—top man in his field."

"What is his field?" Ed asked gloomily.

"Zoology. Whatever these things are in this picture, they're animal, which is his field. Everyone will be sworn to secrecy; you'll be the only one permitted to use a camera, Ed. This could be the biggest thing you'll ever cover, if these things are what I think they are."

"What do you think they are?"

"An unknown species of man—or subman," his boss said, and prudently Ed remained silent. Two months would tell the tale.

But two months didn't tell. Oh, there were plenty of wild rumors by the Nepalese all along the upper route. Hushed stories of the two-legged creature that walked like a man. A monster the Sherpas called yeti. Legends. Strange encounters; drums sounding from snow-swept heights; wild snatches of song drifting down from peaks that were inaccessible to ordinary men. And one concrete fact: a ban, laid on by the Buddhist monks, against the taking of any life in the high Himalayas. What life? Ed wondered.

Stories, legends—but nothing else.

Two months of it. Starting from the tropical flatlands, up through the lush, exotic rain forest, where sun struggled through immense trees festooned with orchids. Two months, moving up into the arid foothills, where foliage abruptly ceased and the rocks and wind took over. Up and ever up, to where the first heavy snow pack lay. And higher still, following the trail laid out by the glider pilot—and what impelled a man, Ed wondered, to soar over Mount Everest in a homemade glider?

Two months, during which Ed had come to dislike Dr. Schenk intensely. Tall, saturnine, smelling strongly of formaldehyde, Schenk classified everything into terms of vertebrate and invertebrate.

So now, standing on this wind-scoured ridge with the shadows falling into the abysses on either side, Ed peered through ice-encrusted goggles, watching Schenk arguing with the guides. He motioned to the ledge above, and obediently the Sherpas moved toward it. Obviously that would be the final camping spot. The two months were over by several days; Schenk was within his rights to call it quits. It was only Ed's assurances that the plateau they were seeking lay just ahead that had kept Schenk from bowing out exactly on the appointed time; that and the burning desire to secure his niche in zoology forever with a new specimen: biped, erect-walking—what?

But the plateau just ahead, and the one after that, and all the rest beyond had proved just as empty as those behind.

A bust. Whatever the unknown creatures were that the glider pilot had photographed, they would remain just that—unknown.

And yet, as Ed slogged slowly up toward where the porters were setting up the bright blue-and-yellow nylon tents, he was nagged by a feeling that that odd-shaped pinnacle ahead looked awfully much like the one in the blurred photograph. With his unfailing memory for pictures, Ed remembered the tall, jagged cone that had cast a black shadow across a snowy plateau, pointing directly toward the little group that was in the center of the picture.

But Schenk wasn't having any more plateaus. He shook his head vehemently, white-daubed lips a grim line on his sun-blistered face. "Last camp, Ed," he said firmly. "We agreed this would be the final plateau. I'm already a week behind schedule. If the monsoons hit us, we could be in serious trouble below. We have to get started back. I know exactly how you feel, but—I'm afraid this is it."

Later that night, while the wind moved ceaselessly, sucking at the tent, they burrowed in sleeping bags, talking.

"There must be some basis of fact to those stories," Ed said to Dr. Schenk. "I've given them a lot of thought. Has it occurred to you that every one of the sightings, the few face-to-face meetings of the natives and these—these unknowns, has generally been just around dawn, and usually when the native was alone?''

Schenk smiled dubiously. ''Whatever this creature may be—and I'm convinced that it's either a species of large bear or one of the great anthropoids—it certainly must keep off the well-traveled routes. There are very few passes through these peaks, of course, and it would be quite simple for them to avoid these locales.''

''But we're not on any known trail,'' Ed said thoughtfully. ''I believe our methods have been all wrong—stringing out a bunch of men, looking for trails in the snow. All we've done is announce our presence to anything with ears for miles around. That glider pilot made no sound; he came on them without warning.''

Ed looked intently at Schenk. ''I'd like to try that peak up ahead—and the plateau beyond.'' When Schenk uttered a protesting cry, Ed said, ''Wait; this time I'll go alone—with just one Sherpa guide. We could leave several hours before daybreak. No equipment other than oxygen, food for one meal—and my cameras, of course. Maintain a strict silence. We could be back before noon. Will you wait long enough for this one last try?'' Schenk hesitated. ''Only a few hours more,'' Ed urged.

Schenk stared at him, then he nodded slowly. ''Agreed. But aren't you forgetting the most important item of all?'' When Ed looked blank, Schenk smiled. ''The gas gun. If you should run across one, we'll need more proof than just your word for it.''

There was very little wind and no moon, but cold, the cold approaching that of outer space, as Ed and one Sherpa porter started away from the sleeping camp, up the shattered floor of an ice river that swept down from the jagged peak ahead.

They moved up, hearing only the squeak of equipment, the peculiar gritty sound of crampons biting into packed snow, an occasional hollow crash of falling ice blocks. To the east already a faint line of gray was visible; daylight was hours away, but at this tremendous height sunrise came early. They moved slowly, the thin air cutting cruelly into their lungs, moving up, up.

They stopped once for hot chocolate from a vacuum bottle, and Ed slapped the Sherpa's shoulder, grinning, pointing ahead to where the jagged peak glowed pink and gold in the first slanting rays of the sun. The Sherpa looked at the peak and quickly shifted his glance to the sky. He gave a long, careful look at the gathering clouds in the east, then muttered something, shaking his head, pointing back, back down to where the camp was hidden in the inky shadows of enormous boulders.

When Ed resumed the climb, the Sherpa removed the long nylon line which had joined them. The route was now comparatively level, on a huge sweeping expanse of snow-covered glacier that flowed about at the base of the peak. The Sherpa, no longer in the lead, began dropping behind as Ed pressed eagerly forward.

The sun was up, and with it the wind began keening again, bitterly sharp, bringing with it a scent of coming snow. In the east, beyond the jagged peak just ahead, the immense escarpment of the Himalayas was lost in approaching clouds. Ed hurried as best he could; it would snow, and soon. He'd have to make better time.

But above, the sky was blue, infinitely blue, and behind, the sun was well up, although the camp was still lost in night below. The peak thrust up ahead, quite near, with what appeared to be a natural pass skirting its flank. Ed made for it. As he circled an upthrust ridge of reddish, rotten rock, he glanced ahead. The plateau spread out before him, gently sloping, a natural amphitheater full of deep, smooth snow, with peaks surrounding it, and the central peak thrusting a long black shadow directly across the center. He paused, glancing back. The Sherpa had stopped, well below him, his face a dark blur, looking up, gesticulating frantically, pointing to the clouds. Ed motioned, then moved around, leaning against the rock, peering ahead.

That great shadow against the snow was certainly similar to the one in the photo—only, of course, the shadow pointed west now, when, later, it would point northwest, as the sun swung to the south. And when it did, most certainly it was the precise—. He sucked in a sharp, lung-piercing breath.

He stared, squinting against the rising wind that seemed to blow from earth's outermost reaches. Three figures stirred slightly, and suddenly leaped into focus, almost perfectly the

camouflaged against the snow and wind-blasted rock. Three figures, not more than a hundred feet below him. Two small, one larger.

He leaned forward, his heart thudding terribly at this twenty-thousand-foot height. A tremor of excitement shook him. My Lord—it was true. They existed. He was looking at what was undeniably a female and two smaller— what? Apes?

They were covered with downy hair, nearly white, resembling nothing so much as tight-fitting leotards. The female was exactly like any woman on earth—except for the hair. No larger than most women, with arms slightly longer, more muscular. Thighs heavier, legs out of proportion to the trunk—shorter. Breasts full and firm. Not apes.

Hardly breathing, Ed squinted, staring, motionless. Not apes. Not standing so erectly. Not with those broad, high brows. Not with the undeniable intelligence of the two young capering about their mother. Not—and seeing this, Ed trembled against the freezing rock—not with the sudden affectionate sweep of the female as she lifted the smaller and pressed it to her breast, smoothing back hair from its face with a motion common to every human mother on earth. A wonderfully tender gesture.

What were they? Less than human? Perhaps. He couldn't be certain, but he thought he heard a faint gurgle of laughter from the female, fondling the small one, and the sound stirred him strangely. Dr. Schenk had assured him that no animal was capable of genuine laughter; only man.

But they laughed, those three, and, hearing it, watching the mother tickling the younger one, watching its delighted squirming, Ed knew that in that marvelous little grouping below, perfectly lighted, perfectly staged, he was privileged to observe one of the earth's most guarded secrets.

He should get started, shooting his pictures; afterward he should stun the group into unconsciousness with the gas gun and then send the Sherpa back down for Dr. Schenk and the others. Clouds were massing, immensities of blue-black. Already the first few flakes of snow, huge and wet, drifted against his face.

But for a long moment more he remained motionless, oddly unwilling to do anything to destroy the harmony, the aching purity of the scene below, so vividly etched in brilliant light and shadow. The female, child slung casually on one hip, stood erect, hand shading her eyes, and Ed grinned. Artless, but perfectly posed. She was looking carefully out and above, scanning the great outcroppings of rock, obviously searching for something.

Then she paused. She was staring directly at him.

Ed froze, even though he knew he was perfectly concealed by the deep shadows of the high cliff behind him. She was still looking directly at him, and then, slowly her hand came up. She waved.

He shivered uncontrollably in the biting wind, trying to remain motionless. The two young ones suddenly began to jump up and down and show every evidence of joy. And suddenly Ed knew.

He turned slowly, very slowly, and with the sensation of a freezing knife plunging deeply into his chest, he saw the male less than five yards away.

He was huge, easily twice the size of the female below. And, crazily, Ed thought of Schenk's little lecture, given what seemed like eons ago, in the incredible tropical grove far below and six weeks before, where rhododendrons grew in wild profusion and enormous butterflies flitted about: "In primitive man," Schenk had said, "as in the great apes today, the male was far larger than the female."

The gas gun was hopelessly out of reach, securely strapped to his shoulder pack. Ed stared, knowing there was absolutely nothing he could do to protect himself before this creature, fully eight feet tall, with arms as big as Ed's own thighs, and eyes (My Lord—blue eyes!) boring into his. There was a light of savage intelligence there—and something else.

The creature made no move against him, and Ed stared at it, breathing rapidly, shallowly and with difficulty, noting with his photographer's eyes the immense chest span, the easy rise and fall of his breathing, the large, square, white teeth, the somber cast of his face. There was long, sandy fur on the shoulders, chest and back, shortening to off-white over the rest of the magnificent torso. Ears rather small and close to the head. Short, thick neck, rising up from the broad shoulders to the back of the head in a straight line. Toes long and definitely prehensile.

They looked intently at each other across the abyss of time and mystery. Man and—what? How long, Ed wondered, had it stood there, observing him? Why hadn't it attacked? Had it been waiting for Ed to make a single threatening gesture—such as pointing a gun or camera? Seeing the calm awareness in those long, slanting, blue eyes, Ed sped a silent prayer of thanks upward; most certainly if he had made a move for camera or gun, that move would have been his last.

They looked at each other through the falling snow, and suddenly there was a perfect instantaneous understanding between them. Ed made an awkward, half-frozen little bow, moving backward. The great creature stood motionless, merely watching, and then Ed did a strange thing: He held out his hands, palms up, gave a wry grin—and ducked quickly around the outcropping of rock and began a plunging, sliding return down the way he'd come. In spite of the harsh, snow-laden wind, bitterly cold, he was perspiring.

Ed glanced back once. Nothing. Only the thickening veil of swift-blowing snow, blanking out the pinnacle, erasing every trace—every proof that anyone, anything, had stood there moments before. Only the snow, only the rocks, only the unending wind-filled silence of the top of the world. Nothing else.

The Sherpa was struggling up to him from below, terribly anxious to get started back; the storm was rising. Without a word they hooked up and began the groping, stumbling descent back to the last camp. They found the camp already broken, Sherpas already moving out.

Schenk paused only long enough to give Ed a questioning look.

What could Ed say? Schenk was a scientist, demanding material proof: If not a corpse, at the very least a photograph. The only photographs Ed had were etched in his mind—not on film. And even if he could persuade Schenk to wait, when the storm cleared, the forewarned giant would be gone. Some farther peak, some remoter plateau would echo to his young ones' laughter.

Feeling not a bit bad about it, Ed gave Schenk a barely perceptible negative nod. Instantly Schenk shrugged, turned and went plunging down, into the thickening snow, back into the world of littler men. Ed trailed behind.

On the arduous trek back, through that first great storm, through the snow line, through the rain forest hot and humid, Ed thought of the giant, back up there where the air was thin and pure.

Who, what was he, and his race? Castaways on this planet, forever marooned, yearning for a distant, never-to-be-reached home?

Or did they date in unbroken descent from the Pleistocene—man's first beginning—when all the races of not-quite-man were giants; unable or unwilling to take the fork in the road that led to smaller, cleverer man, forced to retreat higher and higher, to more remote areas, until finally there was only one corner of earth left to them— the high Himalayas?

Or were he and his kind earth's last reserves: not-yet-men, waiting for the opening of still another chapter in earth's unending mystery story?

Whatever the giant was, his secret was safe with him, Ed thought. For who would believe it—even if he chose to tell?

POSTREADING VOCABULARY EXERCISE

Directions

Beneath the outline of a mountain range are several vocabulary words that describe landforms numbered on the outline. Decide which term best fits the numbered landform and write it on the line below. For any words not used, write a short definition of your own. Refer to your dictionary if necessary.

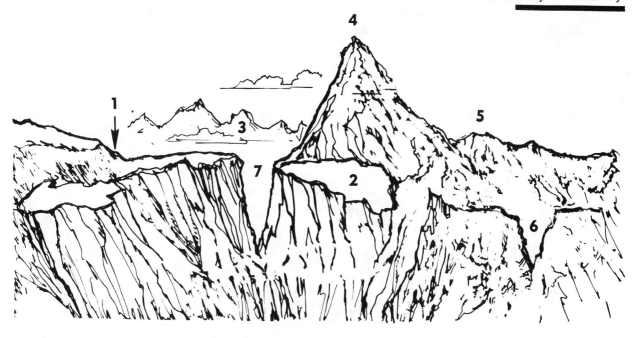

plateau	pinnacle	massif
acclimatized	escarpment	vista
crevass	grandeur	abyss

1. _____   2. _____

3. _____   4. _____

5. _____   6. _____

7. _____

8. _____

9. _____

THINKING ABOUT WHAT YOU'VE READ

1. Although Dr. Schenk was the leader of the expedition, he wanted to end the climb and return home. What were some of the events that led him to make this decision?
2. Why did Ed back away, leaving the area, instead of photographing the family of creatures?
3. Do you think the magazine editor was correct to give Ed the shooting assignment, or should some of the mysterious sightings of strange creatures be left uninvestigated?

COMPREHENSION— recognizing supporting details

Directions

The old saying, "One picture is worth a thousand words," may be true on some occasions. Sometimes, however, it is just as important to be able to describe in words what you have seen. Read the main ideas below from the story and supply the necessary supporting ideas from the list that follows.

1. The expedition appeared to be a failure.

    A. _____

    B. _____

    C. _____

2. The glider pilot's attempt to fly over Mount Everest had been both a success and a failure.

   A. _____

   B. _____

   C. _____

   D. _____

3. Ed's plan to locate and photograph the creatures was a good one.

   A. _____

   B. _____

   C. _____

He planned to leave at daybreak.
The group had found nothing in two months of searching.
He had successfully taken a picture of the creatures.
He planned to travel light—no equipment except for the necessary oxygen tanks.
The pilot spotted the creatures playing at an altitude of 25,000 feet.
Dr. Schenk, the expedition leader, argued with his guides.
He was unable to find the updraft he needed to clear Everest.
He planned to take just one guide with him up the mountain.
They had done nothing but climb the mountain, scrambling over ice and snow.
Soaring over Everest was a goal he had not achieved.

COMPREHENSION— making comparisons

Directions

There were several similarities as well as differences in the way the two men spotted the creatures. Using the story as your guide, decide whether each phrase below reflects a similarity or a difference between Ed and the glider pilot. If it is a similarity, mark both boxes with an **S**. If the action describes something just one of the men did, place an **X** in the box below that man's occupation.

	PILOT	PHOTOGRAPHER
1. Came upon the creatures quietly		
2. Soared over the peak, looking down on the creatures		
3. Worked for a large magazine in New York		
4. Saw a family of the creatures		
5. Was able to photograph the creatures		
6. Described the creatures as tall and able to walk erect		

ACADEMIC SKILLS— using the card catalog

Legends rarely die. Through the centuries, stories and tales have been passed from one generation to the next. Many legends grow with the culture of a particular people and are used to explain their history. The legends that remain today are interesting to listen to but are not generally regarded as an accurate accounting of the past. Many of these legends are concerned with mysterious beasts that are said to live or have lived in regions of the world.

Directions

Using the library card catalog, locate at least two books or articles that discuss the following legendary creatures and then fill in the forms below. Record the location of each of these legendary creatures on the map following the forms. Pick any three creatures to research.

| Yeti | Bigfoot | Loch Ness Monster |
| Phoenix | Unicorns | Champ |

Legend _____

Source: Title _____

    Author _____

    Publisher _____

    Publication Date _____

Source is a ☐ newspaper ☐ magazine ☐ reference ☐ book

Source: Title _____

    Author _____

    Publisher _____

    Publication Date _____

Source is a ☐ newspaper ☐ magazine ☐ reference ☐ book

Legend _____

Source: Title _____

    Author _____

    Publisher _____

    Publication Date _____

Source is a ☐ newspaper ☐ magazine ☐ reference ☐ book

Source: Title _____

    Author _____

    Publisher _____

    Publication Date _____

Source is a ☐ newspaper ☐ magazine ☐ reference ☐ book

Legend _____

Source: Title _____

       Author _____

       Publisher _____

       Publication Date _____

Source is a ☐ newspaper   ☐ magazine   ☐ reference   ☐ book

Source: Title _____

       Author _____

       Publisher _____

       Publication Date _____

Source is a ☐ newspaper   ☐ magazine   ☐ reference   ☐ book

WRITING SKILLS—writing a short report from multiple sources

It is interesting to note that many of the legends that have survived through the centuries do contain some elements of truth. The characters found in legends are often credited with some fantastic achievement, such as forming the Great Lakes or creating violent storms.

Directions

Using the sources you have already located in the previous activity, complete the research outline below and then write a short report that can be presented orally to the class.

Legend: _____

Continent of Origin: _____

What group of people is responsible for the development of the legend? _____
_____

What is the earliest known sighting or visitation of this creature? _____
_____

Is there any historical basis for believing this creature might actually have existed, or exists, today?
_____

(*Develop your own questions and continue on a separate sheet of paper.*)

PART TWO

MACABRE

INTRODUCTION TO PART TWO

Isaac Asimov

It seems to me that students do not often notice the names of the writers of stories. Perhaps they are not even certain that writers really exist, but feel that stories exist in the way that hills do — they simply always have been there.

I have sometimes observed that when parents introduce their children to me as the person who wrote such-and-such a story that they liked so much, the comment seems to elicit an odd response. The children tend to look at me with uncertainty, as though they expect that this is some sort of game being played upon them. I *can't* be the writer; there simply *isn't* any writer!

And yet, some mention of the existence and personality of the writer may increase the readers' interest in the story. It may make them feel that if *someone* can write a story, then someday maybe they can do so. This is not a bad notion for them to have. Although most students will not become professional writers, all of them should *try* to write a story. Not only will such an exercise develop the imagination and, possibly, a few writing skills, but, at the very worst, it will convince them of the difficulty of the task. It will also give them a greater appreciation of the stories they read.

To me, of course, the writers of science fiction are very real. Indeed, I have met many of them and have known some of them for years. Back in the late 1930s, a number of teenagers who were avid science fiction readers got together and called themselves "The Futurians." (If students are told this story, they might profitably speculate on the meaning of the title and its relation to science fiction.) As it happens, this was a most unusual group, for as its members grew older, most of them became very prominent as science fiction writers. I was one of them; another was Richard Wilson, who has three stories in this collection. Frederik Pohl, Cyril Kornbluth, Damon Knight, and James Blish were, at one time or another, members of The Futurians, as were such well-known editors as Donald Wollheim and Robert Lowndes.

Why this should have been, I am not certain. I feel, however, that we stimulated each other; that the interests of one fed upon those of another; that the occasional disappointments of one were buoyed up by the hopes of another; that the occasional triumphs of one cheered all the rest. The common nature of our interests has held us together through many decades, and those of us who are still alive are friends to this day.

Just as writers are sometimes ignored by young readers, so, too, are titles, which also applies to older readers. I frequently find that when people are eager to tell me about a story of mine they liked, they sometimes have a little trouble remembering the title — or, if they do think they remember it, they are quite likely to get it wrong. I suppose if you asked someone why he or she couldn't remember a title, the person might say that, after all, titles aren't important; it's the story that counts. Nevertheless, a title is *part* of a story; and if it is a good one, it *adds* to the story.

Consider "The King of Beasts." Before students read the story, ask them to name the king of beasts. Most are sure to say the lion. The lion is commonly given the title because of its size, its ferocity, the sound of its roar, the calm majesty of its expression, the magnificence of its mane, and because it seems to act like a king. Ask students why the lion is the king of beasts, and most are quite likely to cite one or more of the characteristics I have just mentioned. However, after they have read the story, ask once again who is the king of beasts and why. You should now get a very different answer. In fact, the students will not only be sure to think more deeply about the proper characteristics the animal that rules all the rest should have, but also will, of their own accord, begin to wonder whether human beings are playing their roles properly.

Or consider a title like "The Boy with Five Fingers." To anyone beginning that story, the title might seem silly. Why make a point of having five fingers? Once the story is finished, however, the title doesn't seem silly at all. In fact, it is extremely appropriate and gives a rather grisly point to the story.

"Appointment at Noon" and "The Last Paradox" are both titles that seem mild an unimpressive. Again, however, after the stories are read, their titles seem to gain new and macabre meanings. Their very mildness now makes them seem more cold and menacing. Would the stories have been more effective if they had been entitled "The Coming of Death" and "The Man from Hell"? If students discuss the nature of titles, they would be sure to see that these alternatives would simply spoil the stories by telegraphing and weakening their punch.

THE KING OF BEASTS
Philip José Farmer

Prereading Vocabulary

biologist: one who studies life and life processes. (See next definition for example.)

distinguished: maintaining excellence in life and action; set apart from others as a mark of honor. *(The* biologist *showed the* distinguished *visitor through the laboratory.)*

extinct: no longer living; forever lost. (See next definition for example.)

species: a class of things with similar characteristics and qualities. *(It was the task of the biologist to try and reintroduce many* extinct species.*)*

wanton: cruel, without mercy. (See next definition for example.)

exterminate: to destroy completely. *(The biologist was attempting to bring back those creatures that had been* wanton(ly) exterminate(d).*)*

quaver: to shake; to have part of your body tremble. *(The biologist's voice* quaver(ed) *as he spoke about the contents of the last tank.)*

The biologist was showing the distinguished visitor through the zoo and laboratory.

"Our budget," he said, "is too limited to re-create all known extinct species. So we bring to life only the higher animals, the beautiful ones that were wantonly exterminated. I'm trying, as it were, to make up for brutality and stupidity. You might say that man struck God in the face every time he wiped out a branch of the animal kingdom."

He paused, and they looked across the moats and the force fields. The quagga wheeled and galloped, delight and sun flashing off his flanks. The sea otter poked his humorous whiskers from the water. The gorilla peered from behind bamboo. Passenger pigeons strutted. A rhinoceros trotted like a dainty battleship. With gentle eyes a giraffe looked at them, then resumed eating leaves.

"There's the dodo. Not beautiful but very droll. And very helpless. Come, I'll show you the re-creation itself."

In the great building, they passed between rows of tall and wide tanks. They could see clearly through the windows and the jelly within.

"Those are African Elephant embryos," said the biologist. "We plan to grow a large herd and then release them on the new government preserve."

"You positively radiate," said the distinguished visitor. "You really love the animals, don't you?"

"I love all life."

"Tell me," said the visitor, "where do you get the data for re-creation?"

"Mostly, skeletons and skins from the ancient museums. Excavated books and films that we succeeded in restoring and then translating. Ah, see those huge eggs? The chicks of the giant moa are growing within them. There, almost ready to be taken from the tank, are tiger cubs. They'll be dangerous when grown but will be confined to the preserve."

The visitor stopped before the last of the tanks.

"Just one?" he said. "What is it?"

"Poor little thing," said the biologist, now sad. "It will be so alone. But I shall give it all the love I have."

"Is it so dangerous?" said the visitor. "Worse than elephants, tigers and bears?"

"I had to get special permission to grow this one," said the biologist. His voice quavered.

The visitor stepped sharply back from the tank. He said, "Then it must be... But you wouldn't dare!"

The biologist nodded.

"Yes. It's a man."

POSTREADING VOCABULARY EXERCISE

Directions

Read each of the following sentences. If the vocabulary term is correctly used, place a **C** on the line. If the term is used incorrectly, place an **I** on the line and rewrite the sentence on the lines below so that it makes sense. Refer to your dictionary if necessary.

____ 1. The speaker was so self-controlled and calm that his voice **quavered** as he spoke.

____ 2. It was the obligation of the pest control company to **exterminate** all of the grasshoppers in the botanic gardens.

____ 3. The **biologist** spent all of his time studying the rotation of the Earth and other planets.

____ 4. The con artists were **distinguished** members of the community.

____ 5. The dodo bird has been **extinct** for many decades.

____ 6. The arsonist **wantonly** destroyed the historic building.

____ 7. There are many **species** of fish on the Earth.

_____

_____

_____

_____

_____

_____

_____

_____

THINKING ABOUT WHAT YOU'VE READ

1. What was the biologist's goal in conducting his work?
2. What do you suppose the biologist thought were the major reasons for the extinction of the human race?
3. Why would it be necessary for the biologist to obtain special permission to reintroduce the human race into the environment?

COMPREHENSION— cause and effect

Directions

It's difficult to imagine anyone consciously pushing any species to the edge of extinction. But man has caused several species of lower animals to disappear from the Earth. Indicate the probable effect of the following statements and questions.

1. The biologist continued his work to the best of his ability, given his limited budget.

_____

_____

2. In earlier years, generations of lower animals were wantonly exterminated.

_____

_____

3. The biologist was carefully nurturing the growth of several extinct species.

_____

_____

4. What effect did the contents of the last tank have on the distinguished visitor?

_____

_____

5. What effect would the biologist's work have on the world if his efforts were successful?

_____

_____

COMPREHENSION— summary

Directions

Scientific research is often funded by governments and private foundations. But before organizations will pay for research, they want to know if the research has been thoroughly planned, if it will be carefully conducted, and if the results are likely to benefit mankind and the environment.

Using the information you gathered from the story and the previous exercise, summarize the research being conducted. Make certain to note what is being done and why. Note the benefits we will receive from this research.

_____

_____

_____

_____

_____

_____

_____

_____

_____

_____

_____

_____

_____

_____

_____

_____

_____

STUDY SKILLS— classification

The unintended extermination of animals has become a serious problem on Earth. Each year the list of animals that are considered to be in danger of becoming extinct grows longer. The endangered species range in size from the small snail darter to the giant panda, and do not come just from remote corners of the world. Pollution in the major cities of America is having an effect on animals that have lived in the region for centuries. Without direct action, the endangerment and extinction of animals will continue into the next century.

Directions

Using the library as a resource, divide the following list of animals into three categories: prehistoric, extinct, and endangered. If you wish, you may supplement the lists with the names of other prehistoric, extinct, and endangered animals.

bald eagle	Florida Key deer	passenger pigeon
baluchistan	glyptodant	saber-toothed tiger
brown pelican	great auk	urus (wild ox)
dodo	Japanese sea lion	woolly mammoth
eohippus	moa	woolly spider monkey

PREHISTORIC	EXTINCT	ENDANGERED
1.	1.	1.
2.	2.	2.
3.	3.	3.
4.	4.	4.
5.	5.	5.
6.	6.	6.
7.	7.	7.

WRITING SKILLS— report writing

One of the best ways to reduce the number of animals being placed on the endangered species list is to develop a public awareness of the problem. Millions of dollars are spent each year by such organizations as the National Wildlife Federation in an effort to bring the problem to the attention of the public. If their efforts are to be successful, more must be done to promote animal conservation and protection.

Directions

Using any means you wish (posters, recorded messages, videotaped public service announcements, reports, charts, books, etc.), develop a public awareness campaign aimed at saving one of the endangered species listed in the previous exercise. Use as much factual information as possible. The following outline should help you get started.

What animal are you focusing upon? _____

When was it placed on the endangered species list? _____

What is its natural habitat? _____

What is causing the population of this species to decrease? _____

_____

_____

What can be done to save this species? _____

_____

_____

How many of this species remain? _____

Are there any organizations that are currently working to save this species of animal? If so, how may

they be reached? _____

_____

_____

LITTLE WILLIAM
Patricia Matthews

Prereading Vocabulary

tremendous: very large, expansive. *(Winston Hammersmith was a man of* tremendous *intelligence.)*

unparalleled: never before accomplished; unequaled. *(Winston's feelings of revenge toward Miss Lundy were* unparalleled.*)*

warrant: to guarantee, justify. *(It was Miss Lundy's opinion that she had done nothing to* warrant *Winston's revenge.)*

brood: to spend time thinking unhappy thoughts. *(Winston never mentioned the anger he felt toward Miss Lundy. He chose to* brood *about it in solitude.)*

solitude: being alone; privacy. (See next definition for example.)

beguiling: delightful, charming. *(Miss Lundy found Little William not only friendly and kind, but* beguiling *as well.)*

inflexible: rigid; not able to be changed. *(Miss Lundy's back appeared to be* inflexible *as she rose to her full height.)*

spinster: a woman who remains unmarried into her later years (not a complimentary term). *(Since she never married, Miss Lundy was considered to be a* spinster.*)*

consciousness: the state of being awake and aware. *(Miss Lundy was brought to* consciousness *by Little William's crying.)*

Winston Hammersmith, Ph.D., was a man of tremendous intellect and unparalleled ability; therefore, his revenge upon Miss Leontyne Lundy was far from ordinary. What Miss Lundy did to Mr. Hammersmith to warrant this revenge is really not important; it will suffice that she rejected him in such a manner that she wounded his spirit and ego, and that is a thing that a woman should never do to a man.

Hammersmith did not show his hurt at the time, but he brooded in solitude, and then, one day he came to her door, leading little William by the hand.

Little William was a child to melt the heart of any maiden lady. His sturdy, three-year-old boy was dimpled and golden, his hair, honey floss. His eyes, blue, and round as quarters, were fringed with beguiling, long, dark lashes, and his smile was sweet as morning.

Miss Lundy looked at him, and Hammersmith thought that he saw her expression soften, for the least bit of a moment. She looked at Hammersmith without apparent emotion.

"Good morning, Winston. What can I do for you?"

Hammersmith smiled a smile which he hoped looked apologetic.

"Leontyne," he said, looking deeply into her eyes. "I hesitate to ask you, but I seem to be in a bit of difficulty, and there was no one else to turn to."

Miss Lundy raised her delicate eyebrows. "Do come to the point, Winston. You always draw things out so."

Hammersmith tried the smile again. "To be sure. Well, you see, it's little William, here. He's my nephew. His mother is a widow, quite alone in the world, and at the moment very ill; so she sent him to me."

Miss Lundy's voice was cold. "I fail to see the problem, Winston."

"It's business. Some of my patents. I must go to Washington, at once. It's quite impossible, you know, for me to take a three-year-old child. I was hoping that you ..."

Leontyne smiled a superior smile. "You want me to keep him for you while you're gone. I don't know why you cannot be more direct, Winston. Of course, it is a great deal to ask."

She looked at the golden child out of the corner of her eye and Winston noticed, with pleasure, that her expression softened again. She drew herself up, in the gesture that Winston recognized so well; it made her look ten feet tall.

"Very well," she said. "However, you must understand that I am not doing it for you, but for the child. I trust that you won't be gone too long."

"A week," said Hammersmith hastily. "Only a week."

"Very well, Winston; you may bring him in."

She turned and led the way into the house. Hammersmith followed with the child and a suitcase, smiling a secret smile at Leontyne's straight, slender, inflexible back.

One, two, three days went by in rapid, pleasurable succession. Little William was adorable, an exceptional child. He ate with a spoon and did not soil himself. He was toilet trained. He had many moods, each more charming than the other. He did not cry. He called Miss Lundy "Aunty" and kissed her nicely on the cheek. She was enslaved.

She spent hours with him. She walked with him in the garden, where they picked flowers and played Ring-Around-The-Rosy. Hammer-

smith would not have recognized her. Her eyes sparkled and her cheeks were pink.

The week went by without Miss Lundy even being aware of the passage of the days. When Hammersmith called her from Washington, to tell her that he had been delayed, and to ask if she would keep little William another week, she went limp with joy. So, they began their second week, the cool spinster and the golden boy child, little William.

It was during this second week that William began to change. At first Miss Lundy hardly noticed it. Little things, so small that they slipped by the conscious mind.

On Monday he began to cry, at night, late. Miss Lundy struggled to consciousness and hurried to the side of her lovely child. He quieted at once when she came into the room, but the next night he cried again.

On Tuesday his table manners began to go down hill rapidly. Where before he had neatly inserted the laden spoon into his cherubic pink rosebud of a mouth, now he managed to drop most of the load on his clothes and face. Miss Lundy told herself that this was perfectly natural. Children often regressed briefly. She had read this in a book, she was certain.

Wednesday, it was toilet training; which was almost too much for her maidenly sensibilities. Before, he had always told her, in his little way, when it was necessary for him to use the bathroom facilities. Now ... She wouldn't have admitted it, but her golden boy was becoming tarnished.

On Thursday, he refused to kiss her when she put him down for his nap; not only that, but he began crying, striking at her with his small fists. It almost broke her heart.

On Friday she found him in the garden. The sunlight was making an aureole of his golden curls and his face was wrinkled in concentration. She went up to him to see what he was doing. He stopped for a moment and looked up at her with his beatific, shining smile. She looked down at his chubby pink hands, and gagged. He was happily pulling the legs off of a large grasshopper, which squirmed and moved horribly in the small grasp.

On Saturday, two weeks to the day of his arrival, Leontyne lay in bed late. She felt wan and headachy. She really did hope that Winston would come for the child today. She felt an uncomfortable ambivalence in her feeling for William. He was so pink, so warm, so adorable, but on the other hand ...

She moved the scented cloth that rested upon her forehead, then sat bolt upright as a horrible sound split the air and struck her in the vulnerable spot between her eyes.

She stumbled to her feet and wrenched open the door. There was a flash of color; it took her a moment to make out the form of Daisy, the big yellow cat that belonged to her cook. He looked very strange, all patchy. She looked closer, as he cowered in the corner at the end of the hall. His hair was burned off in uneven patches. At the other end of the hall stood little William, his small face all innocence, his smile like light, his plump little hand clutching a large bunch of kitchen matches.

A strange feeling rose in Miss Lundy's breast. Slowly, like lava mounting, it rose as she walked toward the smiling child. He stood there, looking up at her, as she raised her arm high and brought her hand down across the small face.

She watched from somewhere outside herself, as the small form went tumbling, bouncing, jarring down the long stairway and lay in a crumpled heap at the bottom.

For a painful moment she stood there, staring down, then she found that it was possible to move and went stumbling down to kneel at his side; to smooth the golden hair back from the white forehead and away from the wide, blue eyes; to turn the little head that lay so crooked; to push back the little copper wires and wheels and springs that spilled from the ragged gash in the round white throat, to ...

The little wires and wheels and springs, the wires and wheels and ...

A horrible shrieking filled her ears and her mind. She did not know that it was her own voice.

And far away, in Washington, Winston Hammersmith looked at the date on his calendar, and smiled.

POSTREADING VOCABULARY EXERCISE

Directions

Some words have multiple, or more than one, meaning. The Prereading Vocabulary defined the words as they were used in the story. Each of those words has a second meaning, which is presented below. Read the sentences that follow and fill in the correct word. On the line preceding the sentence, place an *A* or *B*, denoting which definition you used for that word.

tremendous: (A) terrible, fear-instilling; (B) very large, expansive
unparalleled: (A) two lines that are not equal distance from each other; (B) unequaled, a
  great accomplishment.
warrant: (A) to guarantee, justify: (B) a written paper authorizing payment
brood: (A) the children of a family; (B) to spend time thinking unhappy thoughts
solitude: (A) a cell where a prisoner can be kept in isolation; (B) privacy
beguile: (A) to cheat someone, to take property through trickery; (B) to delight and charm
inflexible: (A) rigid, not able to be changed; (B) unwilling to change your opinion, steadfast
spinster: (A) a woman who has remained unmarried throughout her later years; (B) a person
  who spins wool or other material into string or yarn
consciousness: (A) the state of being awake and aware; (B) a concern with one's identity

____ 1. Jack lost _____ after falling from the house.

____ 2. The _____ was able to supply her customers with the
        highest quality yarn.

____ 3. The senator did not allow anyone to speak to him regarding the proposed law. He remained

        _____ in his opinion.

____ 4. The thief managed to _____ the spinster, taking all of her
        valuable antiques.

____ 5. After a long week in the tower, the air traffic controller enjoyed the

        _____ of his isolated mountain cabin.

____ 6. The goose carefully nudged her _____ across the busy
        highway.

____ 7. I cashed the _____ at the bank.

____ 8. The performer walked across the thin rope stretched over the canyon. Her accomplishment

        was _____ .

____ 9. The _____ roll of thunder frightened the new-born puppies.

THINKING ABOUT WHAT YOU'VE READ

1. Little William immediately won the affection of Miss Lundy. What were some of the behaviors he used to accomplish this feat?
2. Little William changed slowly at first, in small ways that one would hardly notice. But, as the second week passed, these changes became more dramatic and more frequent. Describe some of these changes.
3. What emotion do you imagine Winston Hammersmith intended Miss Lundy to feel in the end? What emotion did he likely feel?
4. Why do you think Winston planned his revenge in this manner? Why didn't he do something less complicated?

COMPREHENSION— sequencing

Directions

To Miss Lundy's distress, Winston's scheme for revenge worked perfectly. Little William's program had been carefully developed. When Winston, miles away in Washington, looked at his watch, he knew his spiteful plan had been carried out in a logical, precise sequence. Read the following phrases. Write the initials of the character described in the phrase on the first line preceding the phrase. Indicate the order in which the event took place in the story on the second line.

Winston Hammersmith = WH
Leontyne Lundy = LL
Little William = LW

1. _____ _____ Brooded over Miss Lundy's rejection
2. _____ _____ Lost all table manners
3. _____ _____ Spent the first three days of the visit acting adorable; had many charming moods
4. _____ _____ Fell down the steps, breaking into several pieces
5. _____ _____ Arrived at the house with a child in hand
6. _____ _____ Began crying in the middle of the night
7. _____ _____ Called from Washington to report a delay in returning
8. _____ _____ Left for Washington
9. _____ _____ Stayed late in bed, exhausted from the events of the week
10. _____ _____ Agreed to take the child for a week

COMPREHENSION— summarizing

Directions

Revenge is not considered a legitimate reason to act unpleasantly toward another person. Winston's reputation in the community would likely suffer as a result of his scheme. The complete story of what he had done would be told and retold. Each time it was mentioned, the account of the actual events would change slightly. Using your answers from the previous exercise, write a summary of the story, explaining in some detail what happened. Continue your summary on a separate piece of paper if needed.

_____

_____

_____

_____

_____

_____

ACADEMIC SKILLS— following and writing directions

Building and maintaining a machine as complicated as Little William would certainly take a great deal of skill. Fortunately, it is not always necessary to use such a sophisticated device. A simple, multiple-program robot can usually take care of most needs. You have just been retained by the Hercules Robot Company to write simple, one-sentence assembly instructions that will accompany the diagram below.

Directions

Study the diagram for a few moments. When you are famliar with most of the parts and the meaning of the arrows, write the steps necessary to construct the Hercules PR-1. Each step must be described in one sentence. There can be no more than 25 steps involved in building the PR-1.

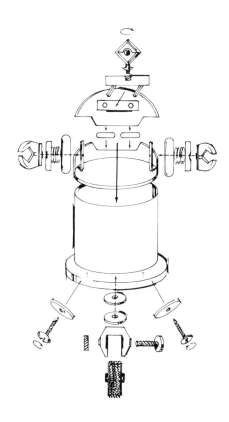

1. _____
2. _____
3. _____
4. _____
5. _____
6. _____
7. _____
8. _____
9. _____
10. _____

(Continue on a separate sheet of paper.)

WRITING SKILLS— writing an owner's manual

Building a robot and learning how to care for one are two distinct skills. Robots, like any complicated instrument, must be well cared for if they are to perform their tasks service-free. In an effort to reduce warranty costs, manufacturers include an owner's manual with a product at the time of sale. An owner's manual includes information about the care, cleaning, and maintenance of the product. It will also give the purchaser trouble-shooting information in the event the machine breaks down.

Directions

On one or more sheets of paper, design and write an owner's manual to accompany the Hercules PR-1. Remember to use as many illustrations as possible.

APPOINTMENT AT NOON
Eric Frank Russell

Prereading Vocabulary

trifler: one who wastes time and energy on small and unimportant matters. *(Henry Curran felt that anyone who took up too much of his time was a* trifler.*)*

alibi: the statement made by a person accused of an act that he was someplace else at the time the act took place; an excuse. *(Henry Curran was a man with an* alibi *for everything. He felt that everyone else made excuses.)*

muse: to think about something at great length; to ponder. *(Curran* muse(d) *over his schedule for the afternoon.)*

precognition: the unexplainable ability to have knowledge of an event before it actually happens. *(The strange man in the outer office demonstrated* precognition *by accurately predicting when Curran would return.)*

martyr: one who suffers as a result of making great sacrifices. *(After Curran shouted at her, she left wearing a* martyr(ed) *expression.)*

sardonic: making fun of something in a scornful manner; mocking. *(Curran* sardonic(ally) *told Miss Reed to throw the blind man down the steps.)*

gaunt: frail, bony. *(The* gaunt *man sat unmoving in the outer office, waiting for his appointment with Henry.)*

tomfoolery: nonsense, silly behavior. *(After their meeting began, Curran hoped the old man was not serious but merely engaged in some sort of* tomfoolery.*)*

Henry Curran was big, busy and impatient of triflers. He had the build of a wrestler, the soul of a tiger, and his time was worth a thousand bucks an hour. He knew of nobody who rated more.
And crime did not pay? Bah!
Jungle tactics paid off. The entire opposition had been conditioned out of men by what is called civilization.

Entering his spacious office with the swift, heavy tread of a large man in fighting trim, Henry slung his hat onto a hook, glanced at the wall clock, noted that it registered ten minutes to twelve.

Planting himself in the seat behind his desk, he kept his expectant gaze upon the door through which he had entered. His wait lasted about ten seconds. Scowling at the thought of it, Curran reached over and thumbed a red stud on his big desk.

"What's wrong with you?" he snapped when Miss Reed came in. "You get worse every day. Old age creeping over you or something?"

She posed, tall, neat, and precise, facing him across the desk, her eyes wearing a touch of humility born of fear. Curran employed only those about whom he knew too much.

"I'm sorry, Mr. Curran, I was—"

"Never mind the alibi. Be faster—or else! Speed's what I like. *Speed*—see?"

"Yes, Mr. Curran."

"Has Lolordo phoned in yet?"

"No, Mr. Curran."

"He should be through by now if everything went all right." He viewed the clock again, tapped irritably on his desk. "If he's made a mess of it and the mouthpiece comes on, tell him to let Lolordo stew. He's in no position to talk, anyway. A spell in jail will teach him not to be stupid."

"Yes, Mr. Curran. There's an old—"

"Shut up till I've finished. If Michaelson calls up and says the *Firefly* got through, ring Voss and tell him without delay! And I mean without delay! That's important!" He mused a moment, finished, "There's that meeting downtown at twelve-twenty. God knows how long it will go on but if they want trouble they can have it aplenty. If anyone asks, you don't know where I am and you don't expect me back before four."

"But, Mr. Curran—"

"You heard what I said. Nobody sees me before four."

"There's a man already here," she got out with a sort of apologetic breathlessness. "He said you have an appointment with him at two minutes to twelve."

"And you fell for a gag like that?" He studied her with open contempt.

"I can only repeat what he said. He seemed quite sincere."

"That's a change," scoffed Curran. "Sincerity in the outer office. He's got the wrong address. Go tell him to spread himself across the tracks."

"I said you were out and didn't know when you would return. He took a seat and said he'd wait because you would be back at ten to twelve."

Involuntarily, both stared at the clock. Curran bent an arm, eyed his wristwatch by way of checking the accuracy of the instrument on the wall.

"That's what the scientific bigbrains would call precognition. I call it a lucky guess. One minute either way would have made him wrong. He ought to back horses." He made a gesture of dismissal. "Push him out—or do I have to get the boys to do it for you?"

"That wouldn't be necessary. He is old and blind."

"I don't give a damn if he's armless and legless—that's *his* tough luck. Give him the rush."

Obediently she left. A few moments later she was back with the martyred air of one compelled to face his wrath.

"I'm terribly sorry, Mr. Curran, but he insists that he has a date with you for two minutes to twelve. He is to see you about a personal matter of major importance."

Curran scowled at the wall. The clock said four minutes to twelve. He spoke with sardonic emphasis.

"I know no blind man and I don't forget appointments. Throw him down the stairs."

She hesitated, standing there wide-eyed. "I'm wondering whether—"

"Out with it!"

"Whether he's been sent to you by someone who'd rather he couldn't identify you by sight."

He thought it over, said, "Could be. You use your brains once in a while. What's his name?"

"He won't say."

"Nor state his business?"

"No."

"H'm! I'll give him two minutes. If he's panhandling for some charity he'll go out through the window. Tell him time is precious and show him in."

She went away, brought back the visitor, gave him a chair. The door closed quietly behind her. The clock said three minutes before the hour.

Curran lounged back and surveyed his caller, finding him tall, gaunt, and white-haired. The oldster's clothes were uniformly black, a deep, somber, solemn black that accentuated the brilliance of the blue, unseeing eyes staring from his colorless face.

Those strange eyes were the other's most noteworthy feature. They held a most curious quality of blank penetration as if somehow they could look *into* the things they could not look *at*. And they were sorry—sorry for what they saw.

For the first time in his life feeling a faint note of alarm, Curran said, "What can I do for you?"

"Nothing," responded the other. "Nothing at all."

His low, organlike voice was pitched at no more than a whisper and with its sounding a queer coldness came over the room. He sat there unmoving and staring at whatever a blind man can see. The coldness increased, became bitter. Curran shivered despite himself. He scowled and got a hold on himself.

"Don't take up my time," advised Curran. "State your business or get to hell out."

"People don't take up time. Time takes up people."

"What the blazes do you mean? Who are you?"

"You know who I am. Every man is a shining sun unto himself until dimmed by his dark companion."

"You're not funny," said Curran, freezing.

"I am never funny."

The tiger light blazed in Curran's eyes as he stood up, placed a thick, firm finger near his desk stud.

"Enough of this tomfoolery! What d'you want?"

Suddenly extending a lengthless, dimensionless arm, Death whispered sadly, "You!"

And took him.

At exactly two minutes to twelve.

POSTREADING VOCABULARY EXERCISE

Directions

Use the definitions of the following statements to solve the crossword puzzle below.

ACROSS

2. one who wastes time on unimportant things
4. "I know what will happen to you tomorrow," the fortune teller said. "I have a special ability."
6. to consider carefully
7. foolishness

DOWN

1. "I didn't do it!" the suspect shouted. I have an _____."
3. mocking
5. excessively thin
6. "He suffered for what he believed in. He was a _____."

THINKING ABOUT WHAT YOU'VE READ

1. Why did Henry Curran finally agree to see the old man?
2. What was Henry's initial attitude toward the stranger? Did this attitude suddenly change? If so, how?
3. Who was the old man? What are some other descriptions that have at one time or another been applied to this character?

COMPREHENSION— character identification/sequence

Directions

Henry Curran considered himself to be a very important man. So important, in fact, that everyone else's time and emotions became secondary to his. Henry never realized that the actions of one person could affect (and be affected by) those of another. Place the following items in the correct sequence by writing the letter preceding each item in the proper box below. When you have finished, you should be able to identify the character who carried out the actions in the story. Write that name in the last box.

A. Told his secretary to throw the man out

B. Told her boss that an old blind man was waiting to see him

C. Finally entered Curran's office at 11:58

D. Asked not to be disturbed until four p. n.

E. Entered his office tossing his hat on the hook

F. Told Curran that men don't take up time; time takes men

G. Refused to leave when told to leave

H. Told the old man to leave

I. Died

J. Agreed to see the man for just two minutes

K. Fearfully entered the private office

L. Demanded the man either state his business or leave immediately

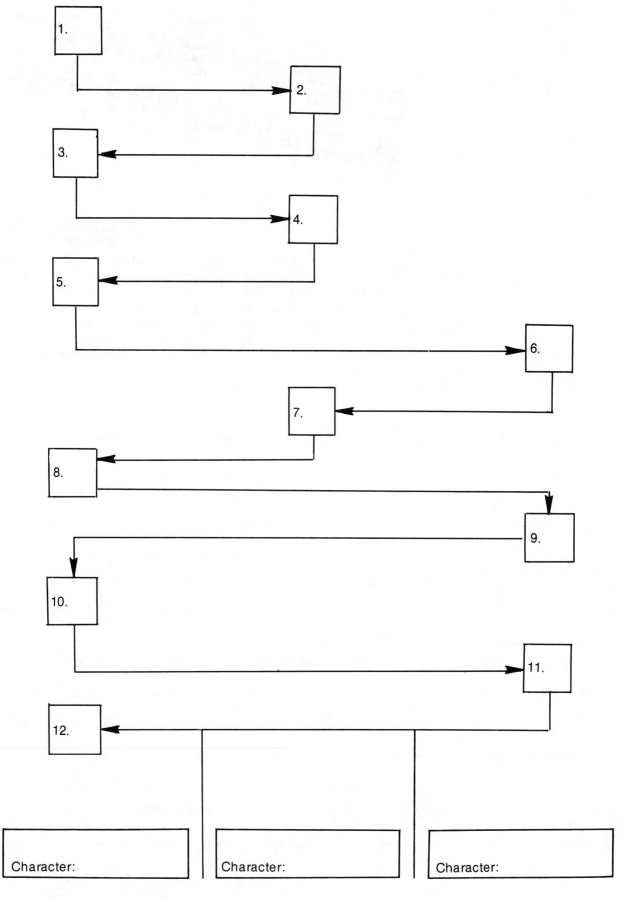

1.

2.

3.

4.

5.

6.

7.

8.

9.

10.

11.

12.

Character:

Character:

Character:

COMPREHENSION— figurative language

Directions

After learning the true nature of the old man's visit, Henry tried to keep himself from falling apart. There was little chance that Henry would actually disassemble. The term "falling apart" is called *figurative language*, designed to give the reader a sense of Henry's mental state. Read the following sentences taken from the story. Explain in your own words what is meant by the figurative language in bold italics.

1. Henry Curran *planted* himself behind his desk.

   _____

   _____

2. The secretary was told to tell the man in the outer office to *spread himself across the tracks*.

   _____

   _____

3. The secretary *gave the man a chair*.

   _____

   _____

4. Mr. Curran told his secretary that if the *mouthpiece* called, to tell him to let Mr. Lolordo *stew* in jail.

   What is a mouthpiece? _____

   _____

   What does it mean to stew in jail? _____

   _____

5. Curran said the only way to succeed in business was to use *jungle tactics*.

   _____

   _____

ORGANIZATION SKILLS—
daily calendar

Whether we appreciate it or not, the world in which we live is becoming busier and more complex. How we spend our days is a mix of obligations and avocations. Obligations are things we **must** do; avocations are hobbies or things that we **like** to do. Henry Curran, as irritating as he was, was able to accomplish what he wanted to do partially because he kept to a tight daily schedule. To achieve everything you want to do in a day, it may be necessary to schedule your time so that you can fit in everything that is important to you.

Directions

Use the weekly calendar below to fill in your schedule for the week. When you have finished, complete the line graph following the schedule.

WEEKLY SCHEDULE					
Time	Monday	Tuesday	Wednesday	Thursday	Friday
6:00 am					
6:30					
7:00					
7:30					
8:00					
8:30					
9:00					
9:30					
10:00					

Time	Monday	Tuesday	Wednesday	Thursday	Friday
10:30					
11:00					
11:30					
12:00 pm					
12:30					
1:00					
1:30					
2:00					
2:30					
3:00					
3:30					
4:00					
4:30					
5:00					
5:30					
6:00					
6:30					
7:00					
7:30					
8:00					
8:30					
9:00					

WEEKLY SCHEDULE

Line Graph

This graph is designed to give you several pieces of information at a single glance. From the preceding schedule, record the total number of hours you spent in each of the activities listed below and mark their locations with points on the graph. These points should then be connected using the following legend:

—————— school/study

══════ recreation

········· eating

- - - - - - - entertainment

M	T	W	Th	F	
					6 hours
					5 hours
					4 hours
					3 hours
					2 hours
					1 hours
					0 hours

WRITING SKILLS— evaluation and proposal

Henry ran his business with an iron hand. He constantly told people what they were going to do and when. Although some people would argue that this type of management is necessary to get the job done, research has shown that people work best when they are happy with what they are doing. In an effort to improve working conditions and organization, people known as efficiency experts study schedules and work plans. They then make suggestions on how the system might be improved. You have just been retained to be your own efficiency expert.

Directions

Review your schedule and line graph to determine what changes and improvements you can make. The efficiency guide below should help you make these decisions. When you have completed the review, write a short proposal on a separate sheet of paper, stating what changes you will make and how they will improve your efficiency.

1. The greatest number of hours in the week are spent _____

_____

2. The least number of hours in the week are spent _____

_____

3. More time should be spent in _____

_____

4. Less time should be spent in _____

_____

5. The most productive time of the day is _____

6. What do you feel is your least productive time during the day (the time of day when you seem

to get nothing done or your mind seems tired)? _____

THE BOY WITH FIVE FINGERS

James Gunn

Prereading Vocabulary

ruins: the remains of something or someplace that has been completely destroyed. *(Scientific investigators found libraries and books among the* ruins *of the Old Race.)*

ancestors: the people from whom one is descended; relatives who lived long before. *(The students were surprised to learn that their* ancestors *had been a part of the Old Race.)*

heap: a collection of things gathered in one place; a pile. *(Miss Harrison told her class that the Old Race lived in houses built on top of one another, as if in a* heap.*)*

flicker: to move unsteadily and rapidly. *(Willie's thin tongue* flicker(ed) *as he stared hatefully at Johnny.)*

discriminate: to act in a positive or negative manner toward someone based on their race, age, sex, or beliefs. *(The Basic Right declared it illegal to* discriminate *against anyone just because they were different.)*

I love Miss Harrison. The other boys laugh at me and say that Miss Davis is prettier or Miss Spencer is nicer. But I don't care. I love Miss Harrison.

Miss Harrison's my teacher. When I grow up we're going to get married. When I tell her that she gets that kind of crinkling around her eye like she does when she's pleased about something, and she says that's fine like she meant it, and I guess she does.

The first time I thought about it was the day Miss Harrison told us about the scientists and the Old Race and the Basic Right. Miss Harrison said we should try to keep track of what the scientists are doing because they are the wisest and maybe if we know more about them we will be wiser, too, and might even be

scientists ourselves some day. But I think what she really wanted to talk about was the Basic Right. Somehow, everyday, she talks about the Basic Right and it must be important because she talks about it so much.

So Miss Harrison said that many, many years ago, before any of us were born, the scientists had uncovered ruins and nobody knew what they were and everybody wondered and thought about them because they were really big.

Somebody said that we had built them long ago and left them and forgotten about them but nobody believed that because we live in little houses far apart and we never had built anything as big as the ruins and never had wanted to build anything like that.

Then somebody else said that the ruins had been built by a race that lived on Earth before we did and had died or something because maybe conditions got different or maybe they went to live on another planet. And everybody said that must be right, so they started calling them the Old Race but nobody knew what they looked like, or did, or anything except that they built these huge places and then went away.

Nobody knew any more than that for years and years, Miss Harrison said, until just a year or so ago when the scientists dug up a place that wasn't all in ruins and found statues and pictures and books and everything. So everybody was all excited and worked on them awfully hard until they could tell what the Old Race was like and just about what was in the books.

Miss Harrison kind of stopped here and looked at us like she does when she's going to tell us something important and we should all get real quiet and listen carefully so we wouldn't miss anything.

Then she said that they had just released the news and the Old Race wasn't really different after all but sort of like ancestors of ours only far away. She said that in lots of ways they were like us only strange and did strange things, and she said we should be sorry for them and glad, too, because maybe if they hadn't been strange we wouldn't be here. Then she told us how strange they were, and I was glad I didn't live then and that I was living now and I was in Miss Harrison's class and listening to her tell us about the Old Race.

Most all of them lived together in these big places, she said, like ants in an ant heap. Everybody gasped at that because we all liked lots of room. But the strangest thing of all, Miss Harrison said, was something else. She stopped again and we all got real quiet. They were all, she said slowly, exactly alike.

Nobody said anything for a moment and then Willie began to laugh the way he does, sort of half-hissing, and pretty soon we were all laughing and Miss Harrison, too. They all had two eyes, she said, and one nose, and one mouth, and two ears, and two arms, and two legs. After every one of those things Willie began hissing again and we all had to laugh. And, Miss Harrison said, they were all stuck in exactly the same place. Their arms and legs all had bones in them that had joints in the middle and at each end.

Though they were all exactly alike, Miss

Harrison said, they thought they could see differences and because of this they did all sorts of strange things until they did the strangest thing of all and ruined all their big places and their children weren't all alike any more. So it went on like that until nobody was alike and here we are. So they were kind of ancestors, like Miss Harrison said.

Then Miss Harrison stopped again and got up slow, the way she does when she wants to make sure everybody will pay attention. We all held our breath. In this room, she said, right now, we have a member of the Old Race.

Everybody let out his breath all at once. We all looked at her but she laughed and said no, she wasn't it. Johnny, she said, stand up, and I stood up. There, said Miss Harrison, is what the Old Race looked like. Everybody stared at me and I felt kind of cold and lonely all at once. Of course, she said, I don't mean Johnny is really one of the Old Race but he looks just like they used to and he even has five fingers on each hand.

All at once I felt ashamed. I put my hands behind me where nobody could see.

Willie started hissing again, but he wasn't laughing now and his thin forked tongue was flickering at me. Everybody moved as far away from me as they could get and started making nasty sounds. If I had been a little younger I might have started to cry, but I just stood there and wished I had a mouth and tongue like Willie's, or a cart like Louise's instead of legs, or arms like Joan's, or fingers like Mike's.

But Miss Harrison stood up straight and frowned, like she does when she's real mad about something and she said she was very surprised and it would seem like everything she'd said had been wasted. Pretty soon everybody quieted down and listened so she wouldn't be mad and she said it looked like what she'd said about the Basic Right hadn't done one bit of good.

Everybody has a right to be different, that was the Basic Right, she said, the foundation of everything and we wouldn't be here now if it weren't for that. And the law says that no one shall discriminate against anyone else because they are different, and that applied to being the same, too. And Miss Harrison said a lot more things I don't remember because I was sort of excited and warm inside. And finally she said she hoped we'd learned a lesson because the Old Race hadn't, and look where they were.

It was right after that I decided I loved Miss Harrison. The other boys say she should have a neck, like Miss Davis, but I don't see why. They say she should have two eyes like me or three like Miss Spencer, but I like her just the way she is and everything she does, like the way she wraps her arm around the chalk when she draws on the board. But I've already said it. I love Miss Harrison.

When I grow up we're going to get married. I've thought of lots of reasons why we should but there's one that's better than any of them.

Miss Harrison and me—I guess we're more different than anybody.

POSTREADING VOCABULARY EXERCISE

Directions

Cross out the one word that does not belong.

1. *Discriminate*: judge, act, poison, evaluate
2. *Flicker*: move, waver, wiggle, fall
3. *Heap*: mound, barren, hill, pile
4. *Ancestor*: forefather, relative, baby, kin
5. *Ruins*: mansion, rubble, wreck, hovel

THINKING ABOUT WHAT YOU'VE READ

1. What was the Basic Right? Why was it so important in this society?
2. Why did Willie begin hissing when Miss Harrison said that Johnny was just like a member of the Old Race?
3. How do you think Johnny felt when he learned that he looked like a member of the Old Race? How would you feel?
4. Do you think that Miss Harrison handled the situation properly?

COMPREHENSION— comparison

Directions

Everyone living in the "new" world had the right to be different — that was the Basic Right of society. It is inevitable that when everyone is different, comparisons will be made. Decide whether the following statements describe something that took place under the Old Race or the New Order, and place an **X** on one of the two lines preceding the statement. If the statement describes a condition that applies to both societies, place an **X** on **both** lines.

OLD RACE	NEW ORDER	
_____	_____	1. This society respected books and learning.
_____	_____	2. This society lived close together in tall buildings.
_____	_____	3. This society originally believed their ancestors had traveled to a distant planet.
_____	_____	4. No two people were the same in this society.
_____	_____	5. This society lived in small buildings with a great deal of distance between neighbors.

COMPREHENSION— fact and opinion

Directions

Occasionally, archeologists — scientists who study the evidence of life in earlier times — will discover a building or design for which they can offer no explanation. The huge, rough stones of Stonehenge in England are an example of such a discovery. Since no one has ever been able to explain why the stones were placed there, scientists can only offer their opinions. Read each of the sentences below and decide if the statement is a fact or an opinion. Place an *F* for fact or an *O* for opinion on the line preceding the statement. Then answer the questions following the statements.

_____ 1. Scientists living in Johnny's day discovered ruins of the Old Race.

_____ 2. The Old Race had always fought among themselves and deserved to be destroyed.

_____ 3. Scientists discovered that the people of the Old Race lived in large communities.

_____ 4. Living together in heaps, as did the Old Race, is not as effective as living in wide-open spaces with few neighbors.

_____ 5. Life is better when nobody looks at all the same.

_____ 6. The people living in Johnny's day all looked different from each other.

_____ 7. The people of the Old Race wanted to destroy their cities.

8. What do you think happened to members of the Old Race?

_____

_____

9. What evidence do you have to support this belief?

_____

_____

10. If you could choose between living in a society where everyone looked basically the same and one in which everyone looked completely different, which would you choose? Why?

_____

_____

_____

_____

_____

STUDY SKILLS— observation/graphing

Miss Harrison and her students looked upon the Old Race as being odd — perhaps in some ways inferior. People of later generations will study the twentieth century and find some of our traditions and behaviors strange, just as we find some of the behaviors of our ancestors to be unusual. Humans, although different in many ways (color of hair, eyes, skin; weight and height; etc.) often try to make themselves appear like each other (style of hair, make-up, clothes, language, etc.). Every few months, a new fad will sweep the country and many people will change some part of their routine to conform with whatever the new fashion is.

Directions

Find a place in your school or neighborhood where you can observe a variety of people. Using the observation form below, note as many similarities among the people as possible. Such similarities might include style and color of dress as well as speech patterns and dialects. Also take note of the obvious differences among people (for example, age, height, and weight). Record these observations by using a simple hash-mark counting system (卌). A 15- to 20-minute observation should give you all the information you need to complete the bar graph that follows the observation form below.

Location of Observations: _____

Date: _____

Time: _____

SIMILARITIES	RECORD HASH MARKS ON LINE
Clothes: Colors	
_____	_____
_____	_____
_____	_____
_____	_____
_____	_____
_____	_____

Shoes:
 Style

_____   _____

_____   _____

_____   _____

_____   _____

_____ :

_____   _____

_____   _____

_____   _____

_____   _____

Directions

Now that you have completed your observation, you can transfer the information to the bar graph below. First, label each of the columns according to your major observation headings. Column A, for example, might be labeled "Clothes — Red"; column B, "Clothes — Blue"; and so on. Each category noted during the observation should be so designated on the lettered lines at the bottom of the bar graph. Next, transfer the counting (hash) marks to the graph. For each hash mark, blacken in one of the boxes above the line on which you have written the category. For example, if you saw 10 people wearing red shirts, you would darken 10 of the boxes above the line on which you had written "Shirts — Red."

11										
10										
9										
8										
7										
6										
5										
4										
3										
2										
1										

A   B   C   D   E   F   G   H   I   J   K

WRITING SKILLS— description

Historians are not the only ones who study twentieth-century lifestyles. Each year millions of dollars are spent by market research firms in an attempt to determine what people will want to buy in the coming months. Unlike Miss Harrison's class, many marketers are happy when people try to be as similar as possible. You have just been retained by the Miracle Marketing Corporation to study the average American student. It is your job to observe how they dress, talk, and what things they like to do with their free time.

Directions

Using the information you gathered through your observations in the previous exercise, as well as what you already know about students, write a fairly detailed profile (on a separate sheet of paper) of a typical student, describing that person in clear yet colorful words. The outline below will help you collect your thoughts before writing the summary report.

STUDENT PROFILE

Students' favorite free-time activities:

_____   _____

_____   _____

_____   _____

Students' favorite foods:

_____   _____

_____   _____

_____   _____

Students' often-used vocabulary:

_____   _____

_____   _____

_____   _____

THE LAST PARADOX
Edward D. Hoch

Prereading Vocabulary

inherent: an absolutely necessary part of something; intrinsic. *(The professor knew there must be a solution to the paradox* inherent *in time travel.)*

gesture: to move your hand (or other limbs) as a sign of expression; a motion. *(Professor Fordley* gesture(d) *as he spoke about his time machine.)*

portray: to draw a picture of something (can also mean to describe something in words). *(John Comptoss was disappointed to learn that actual time travel was not as it had been* portray(ed) *in science fiction.)*

condense: to collect; become thick. *(The water vapor then began to* condense *on the sides of the chamber.)*

vapor: the gaseous state of a substance. *(The water* vapor *collected against the glass dome of the time chamber.)*

vibrate: to move back and forth very rapidly. *(The machine* vibrate(d) *as John Comptoss began his journey into his future.)*

occupant: the person who resides in a certain place; an inhabitant. *(The weight of the human* occupant *was almost too great for the delicate operations of the machine.)*

auxiliary: providing assistance; acting as a secondary or backup. *(The* auxiliary *instruments were functioning perfectly.)*

"It's too bad that G.K. Chesterton never wrote a time-travel story," Professor Fordley lamented as he made the final careful adjustments on his great glass-domed machine. "He, for one, would certainly have realized the solution to the paradox inherent in all travel to the past or future."

John Comptoss, who in a few moments would become the first such traveler outside the pages of fiction, braced the straps of his specially designed pressure suit. "You mean there is a solution? You don't think I'm going to end up in the year 2000 and be able to return with all sorts of fascinating data?"

Fordley shook his head sadly. "Of course not, my boy. I didn't tell you before, because I didn't want to alarm you, but when you step out of my time machine you will not be in the year 2000."

"But . . . but that's what it's set for, isn't it?"

Fordley gestured at the dials. "Certainly it's set for thirty-five years in the future, but there is one slight fact that all the writers about time travel have overlooked till now."

John Comptoss looked unhappy. "What's that, Professor? You think I'll come out in the middle of the Cobalt War or something?"

"It's not that. It's rather . . . well, why have these writers always assumed that travel to the past or future was possible, anyway? We know now that we can—in this machine—increase or decrease the age of an animal, in much the same manner that the age of a traveler through space would change as he approached the speed of light."

"Of course, Professor. We've done it with rocks and plants, and even mice . . ."

Fordley smiled. "In other words, everything that goes into the machine is affected. But what no one ever realized before that *only* the material in the time machine can grow older or younger. When you step out, *you* will be older, but the world will be unchanged."

"You mean the only way we could advance to the year 2000 would be to build a time machine large enough for the entire earth?" John Comptoss asked incredulously.

"Exactly," Fordley replied. "And of course that is impossible. Therefore, time travel as portrayed in fiction will never come to pass."

"So you're going to stick me inside this crazy machine and make me older? Just that and nothing more?"

"Isn't that enough, John? You're twenty-eight years old now—and in a moment you'll be thirty-five years older. You'll be sixty-three . . ."

"Can you bring me back all right? Back to twenty-eight?"

Fordley chuckled. "Of course, my boy. But you must remember everything that happens to you. Everything. There's always a possibility my movie cameras will miss something."

The young man sighed. "Let's get it over with. The whole thing's sort of a letdown now that I'm not going to end up in 2000."

"Step inside," Fordley said quietly, "and . . . good luck."

"Thanks." The heavy door clanged shut behind him, and immediately the condensing water vapor began misting over the glass dome.

Professor Fordley stepped into his control dial and checked the setting. Yes, thirty-five years into the future . . . Not the future of the world, but only the future of John Comptoss . . .

The big machine vibrated a bit, as if sighing at the overload of a human occupant. It took nearly ten minutes before the indicator came level with the thirty-five year mark, and then Fordley flipped the reverse switch.

While he waited for the time traveler to return, he checked the cameras and the dials and the hundreds of auxiliary instruments that had been so necessary to it all. Yes, they were all functioning. He had done it; he had done it with a human being. . . .

The green light above the board flashed on, and he stepped to the heavy steel door. This was the moment, the moment of supreme triumph.

The door opened, slowly, and the blurred figure of John Comptoss stepped out through the smoke.

"John! John, my boy! You're all right!"

"No, Professor," the voice from the steam answered him, sounding somehow strange. "You picked the wrong man for your test. The wrong man : . ."

"What's happened to you, John? Let me see your face!"

"Professor, I died at the age of sixty . . . And there's one place from which even your machine couldn't return me. One place where there is no time . . ."

And then the smoke cleared a bit, and Professor Fordley looked into his face . . .

And screamed . . .

POSTREADING VOCABULARY EXERCISE

Directions

Read the following descriptions and match each
with the appropriate word in the Prereading Vocabulary.

_____ 1. Simone spotted the cab and began waving her hand.

_____ 2. "I cannot play this part!" the actor fumed. He was
having trouble with his role in the play.

_____ 3. The rain fell against the hot steel roof, sending steam
into the afternoon air.

_____ 4. As the chemist applied heat, the substance became
thick and murky.

_____ 5. The guitar strings continued moving back and forth
after they were plucked.

_____ 6. Advertisements sent through the mail are often
addressed to an unknown resident rather than
to a specific person.

_____ 7. The fact that wheels are round is an absolutely
necessary part of their design.

_____ 8. Most aircraft are built with two sets of fuel tanks.
The second set is not always used. It is there
for emergency use.

THINKING ABOUT WHAT YOU'VE READ

1. What was the basic problem with time travel that was never discussed in science fiction writing?
2. Can you see any use for a time machine such as the one Professor Fordley designed and built?
3. What do you suppose were some of John Comptoss' fears just before he became the first
time traveler?

COMPREHENSION— supporting details

Directions

Details determine the quality of an object or event. If John Comptoss had realized what actually was involved in time travel, he may not have taken the job. Following the three main ideas listed below are several statements grouped together that support one of the main ideas. Select the best main idea for each group and write it on the line above the supporting statements. One of the supporting statements in each group does not belong. Draw a line through that statement.

MAIN IDEAS

Time travel was not what people had expected.
John Comptoss prepared to become the first time traveler outside the pages of fiction.
Professor Fordley had spent much time developing the machinery for his time-travel experiment.

1. _____
   A. Wore a specially designed pressure suit.
   B. Stepped inside the time-travel machine.
   C. Was prepared to take careful notes about the trip.
   D. Had read H.G. Wells' book, *The Time Machine*.

2. _____
   A. Had often dreamed of time travel into the future.
   B. Made the final adjustments on the great glass dome.
   C. Carefully constructed the machine.
   D. Placed cameras and recording devices inside the machine to capture the event.

3. _____
   A. There was a paradox inherent in all time travel.
   B. Time did not move forward; only the object inside the machine did.
   C. The only way to travel to the year 2000 would be to put the entire Earth inside the machine.
   D. The Cobalt War would certainly block all travel into the future.

COMPREHENSION— data recall/ character emotion

Directions

If John Comptoss had awakened eager for the day's events, it's safe to assume that his emotions would have changed by nightfall. People's emotions depend upon what is happening around them and how they react to it. Read the following excerpts from the story. Then, in your own words, describe the character's emotions and explain why he might have felt the way he did.

1. "It's too bad that G.K. Chesterton never wrote a time-travel story," Professor Fordley lamented as he made the final adjustments on his great glass-domed machine. "He, for one, would certainly have realized the solution to the paradox inherent in all travel to the past or future."

   A. How did Fordley feel about the fact that Chesterton had never written a story about time travel?

   ☐ disappointed   ☐ glad   ☐ indifferent

   B. Why would the Professor even care whether or not such a story had been written?

   _____

   _____

2. "You mean there is a solution?" Comptoss asked. "You don't think I am going to end up in the year 2000 and be able to return with all sorts of fascinating data?"

   Fordley shook his head sadly. "Of course not, my boy. I didn't want to tell you before, because I didn't want to alarm you, but when you step out of my time machine you will not be in the year 2000."

   "But, that's what it's set for, isn't it?" inquired Comptoss.

   A. Why did the Professor shake his head sadly as he spoke? _____

   _____

   _____

   B. How do you think Comptoss felt at this point?

   ☐ confused   ☐ angry   ☐ eager

   C. Why would he feel this way?

   _____

   _____

3. The young man sighed. "Let's get it over with. The whole thing's sort of a letdown now that I'm not going to end up in the year 2000."

"Step inside," Fordley said quietly, "and . . . good luck."

"Thanks." The heavy door clanged shut behind him, and immediately the condensing water vapor began misting over the glass dome.

A. Why did Comptoss sigh just when he was about to become the world's first time traveler?

_____

_____

B. How would you have felt if you were in his position?

_____

_____

4. The door opened slowly, and the blurred figure of John Comptoss stepped out through the smoke.

"John! John, my boy! You're all right!"

"No, Professor," the voice from the steam answered him, sounding somewhat strange. "You picked the wrong man for your test. The wrong man . . ."

"What's happened to you, John? Let me see your face!"

"Professor, I died at the age of sixty . . . and there's one place from which even your machine couldn't return me. One place where there is no time . . ."

And then the smoke cleared a bit, and Professor Fordley looked into his face . . .

And screamed . . .

A. What emotions do you suppose the Professor felt as he opened the door?

_____

_____

B. What emotions do you suppose he felt as he stared into John Comptoss' face?

_____

_____

C. What emotions do you suppose John Comptoss felt after his trip?

_____

_____

ACADEMIC SKILLS— time line

Time, like life, constantly moves forward. It cannot stand still, as Comptoss discovered. The quality and direction of our lives today actually began taking form generations ago. The disposal of dangerous waste products in the first half of the twentieth century affects us today. Similarly, our actions today will affect the direction and quality of lives many years from now. Changes in society can be marked by using a time line — a means by which historical events can be listed in chronological order.

Directions

Select one of the topics below and complete a time line by listing major events in that topic that occurred over the past 80 years. The statement following the topics will help you get started.

sports	entertainment	education
politics	transportation	

List as many significant events as you can find that pertain to your topic. List the dates as well.

_____   _____

_____   _____

_____   _____

_____   _____

_____   _____

Transfer this information to the time line below.

1900 —

1905 —

1910 —

1915 —

1920 —

1925 —

1930 —

1935 —

1940 —

1945 —

1950 —

1955 —

1960 —

1965 —

1970 —

1975 —

1980 —

1985 —

1990 —

1995 —

2000 —

2005 —

2010 —

2015 —

WRITING SKILLS— prediction

If we successfully solve the paradox inherent in all time travel, it will be possible to see exactly what affect our actions today will have on our descendants in the future. You have just been retained by Chrono-Travel of Greater North America to be the travel coordinator for their new tours into the future.

Directions

Continue the time line into the future, using the same topic you selected in the previous exercise. Then, using a separate sheet of paper, create a travel brochure that describes your trip to the future.

PART THREE

FANTASY

INTRODUCTION TO PART THREE

Isaac Asimov

There are bound to be some stories that will elicit the response, "I don't get it!" This is by no means a reaction that is peculiar to young people alone. I myself, at my advanced age and with my high IQ, sometimes finish a story and respond in exactly that fashion. It isn't really a disgrace. If the writer has tried to be subtle, he or she might just have managed to miss you; or the writer may have been indulging in a private joke that was a trifle too private.

Such perplexity, however, can be turned to good use. In any class, if there seems to be a certain puzzling uncertainty over a story, ask students what they *think* it means. Very likely, there may be several different explanations, and it would be a good exercise indeed to have students defend their accounts and, if possible, support them with quotations from the story.

Consider my story "Dreamworld." We've probably all seen (or read about) monster movies in which the menaces are giant gorillas, giant crabs, giant spiders, or giant dinosaurs. When I was young, there were few movies of this sort, but there were a number of stories in magazines that dealt with the danger of giant insects. In particular, the authors seemed to fear the attack of giant ants. It is this that "Dreamworld" makes fun of, for the young hero dreams he is trapped in a world of giant aunts. Readers either see the aunts/ants play on words at once, or they don't see it at all.

There is this catch, too. In Brooklyn, where I was brought up, the two words are pronounced exactly alike. In other parts of the country, or in social circles other than my own, they may not be.

"The Fun They Had" and "Hometown" are, each in its own way, about nostalgia. In the former story, a little girl longs for an old-fashioned school she has never really experienced; in the latter, a grown woman longs for a city on another world that, presumably, is like the one she once lived in. A great many adults in these mobile times feel the emotions stirred up by such stories. Many Californians were brought up in New York; many New Yorkers were brought up in Iowa. Many suburbanites remember childhoods in crowded city neighborhoods, and many city dwellers remember small towns. In every case, one cannot escape memories and longings for what seems a simpler, warmer time.

But do youngsters experience nostalgia, especially if they have lived more or less in one place all their short lives? Can they imagine nostalgia if they haven't experienced it directly; can they empathize with it? Through these stories there is an excellent opportunity to probe the potentiality of emotion. If *you* lived on the moon, would *you* miss the Earth? Supposed the moon were a newer, neater world, with all its machinery functioning properly. Would you then miss an Earth that might be a great deal dirtier, noisier, and out of order? For that matter, if your education involved home study with an all-knowing computer, would you miss the kind of school you go to now? Is the school you are in really lots of fun? Why?

There is nothing like science fiction and fantasy to get you out of yourself, out of your space, out of your time, and to force new perspectives on you. On the other hand, there is nothing like science fiction to get you out of the ethereal realm of theory and into the nitty-gritty of everyday life.

One of the overwhelming problems in the world today is the threat of nuclear war. Both the United States and the Soviet Union have nuclear arsenals bulging with a variety of weapons that are more deadly than anything the world ever saw prior to 1945. Each country can unleash incredible destruction upon the other. What *power*!

The catch is this: if either country launches a nuclear strike, it probably won't be able to prevent the other from retaliating at once. In less than an hour, dozens of nuclear bombs may explode on the soil of each country, and both countries will be utterly devastated. What's more, the dust thrown into the atmosphere and the radioactive fallout may render the Earth completely unliveable and destroy all of humanity, including the populations of those countries which were in no way concerned with the quarrels of the super-powers.

Thus, unless they wish to be destroyed, neither the United States nor the Soviet Union can start a nuclear war. But each must live in terror that, for some reason, the other will start one. In reality, however, the two nations have no power at all; they have only the *illusion* of power.

But why should they want power that is only an illusion? It seems insane. Perhaps, though, that is the way people are. Does the owner of the big car in "Speed of the Cheetah, Roar of the Lion" have power, or only the illusion of power? If he has only the illusion, is he satisfied? How does that apply to the nuclear stalemate between the United States and the Soviet Union?

DREAMWORLD
Isaac Asimov

Prereading Vocabulary

devotee: an enthusiastic follower of a person or belief. *(Edward Keller had been a science fiction* devotee *for several years.)*

pious: dedicated, religious. *(Edward's aunt held a* pious *memory for her deceased sister.)*

deceased: dead. *(Since his mother was* deceased, *Edward lived with his aunt.)*

waver: to move back and forth; to be unable or unwilling to make up your mind. *(Edward's aunt* waver(ed) *between toleration and exasperation.)*

tolerance: being able to listen to and respect others; making allowances for differences in people. *(Aunt Clara sometimes showed* tolerance *for Edward's interest in science fiction.)*

exasperation: anger or annoyance because of another person's behavior. *(Aunt Clara frequently demonstrated her* exasperation *with Edward's dreams.)*

severe: strict or harsh treatment of another. *(Aunt Clara often spoke to Edward in a* severe *tone of voice.)*

coalesce: be permanently joined together. *(Hundreds of voices* coalesce(d) *into a thunderous roar.)*

A t thirteen, Edward Keller had been a science fiction devotee for four years. He bubbled with galactic enthusiasm.

His Aunt Clara, who had brought him up by rule and rod in pious memory of her deceased sister, wavered between toleration and exasperation. It appalled her to watch him grow so immersed in fantasy.

"Face reality, Eddie," she would say, angrily.

He would nod, but go on, "And I dreamed Martians were chasing me, see? I had a special death ray, but the atomic power unit was pretty low and—"

Every other breakfast consisted of eggs, toast, milk, and some such dream.

Aunt Clara said, severely, "Now, Eddie, one of these nights you won't be able to wake up out of your dream. You'll be trapped! Then what?"

She lowered her angular face close to his and glared.

Eddie was strangely impressed by his aunt's warning. He lay in bed, staring into the darkness. He wouldn't like to be trapped in a dream. It was always nice to wake up before it was too late. Like the time the dinosaurs were after him—

Suddenly he was out of bed, out of the house, out on the lawn, and he knew it was another dream.

The thought was broken by a vague thunder and a shadow that blotted the sun. He looked upward in astonishment and he could make out the human face that touched the clouds.

It was his Aunt Clara! Monstrously tall, she

bent toward him in admonition, mastlike forefinger upraised, voice too guttural to be made out.

Eddie turned and ran in panic. Another Aunt Clara monster loomed up before him, voice rumbling.

He turned again, stumbling, panting, heading outward, outward.

He reached the top of the hill and stopped in horror. Off in the distance a hundred towering Aunt Claras were marching by. As the column passed, each line of Aunt Claras turned their heads sharply toward him and the thunderous bass rumbling coalesced into words:

"Face reality, Eddie. Face reality, Eddie."

Eddie threw himself sobbing to the ground. Please wake up, he begged himself. Don't be caught in this dream.

For unless he woke up, the worst science-fictional doom of all would have overtaken him. He would be trapped, *trapped*, in a world of giant aunts.

POSTREADING VOCABULARY EXERCISE

Directions

The words we read and write bring visual images to mind. In this exercise, the visual images in the first column below refer to the sound of the word. First, decipher the puzzle and write the vocabulary term on the line under the illustration. Then, match the term with the proper letter of the definition in the second column. Refer to the Prereading Vocabulary if you need help.

A. Dedicated, religious

B. To move back and forth

C. To be permanently joined

D. To treat another in a harsh manner

E. An enthusiastic follower

F. To listen to and respect the opinions of others.

G. To become angry with another

H. Dead

1. _____

2. _____

D + BALTIC RED NORTH 〜〜〜〜 + D =

3. _____

4. _____

5. _____

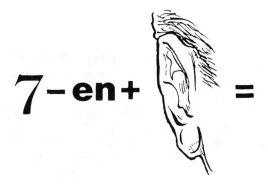

6. _____

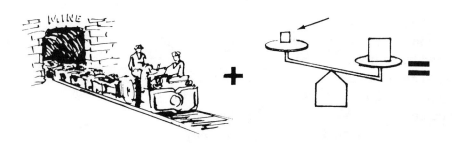

7. _____

8. _____

THINKING ABOUT WHAT YOU'VE READ

1. What is the difference between reality and fantasy?
2. Why was Aunt Clara concerned about Edward's fantasies?
3. How might she more effectively convince her nephew to spend more time on realistic thoughts?
4. Edward considered what it would be like to be trapped in a dream. What do you imagine are some of the things he feared?

COMPREHENSION— distinguishing between fantasy and reality

Directions

Eddie should have listened to his Aunt Clara.
It wasn't long before he wasn't sure whether he was living his reality or
a fantasy. Read the following sentences and decide whether they describe
what Eddie should have recognized as fantasy (label with an **F** on the line)
or his reality (label with an **R**.)

____ 1. Eddie was fascinated by science fiction.

____ 2. Eddie was surrounded by monsters.

____ 3. Eddie's mother was no longer living.

____ 4. Every other breakfast consisted of eggs, toast, and milk.

____ 5. Eddie was being chased by Martians.

____ 6. Eddie did have a special death ray, but the atomic power unit was too low to make it useful.

____ 7. Eddie had once been pursued by dinosaurs.

____ 8. Eddie usually woke up before his dreams became too frightening.

____ 9. Aunt Clara warned Eddie that he should start facing his reality.

____ 10. Eddie finally found himself being circled by an army of giant aunts.

COMPREHENSION— figurative language

Directions

Eddie's morning diet of breakfast and recounted
dreams concerned his aunt. He would sit at the table describing his horrors
in great detail, letting his words paint a ghastly picture. Read the following
sentences and explain in your own words the meaning of the word or phrase
in bold italics.

1. Edward **bubbled with galactic enthusiasm** when he spoke about science fiction.

   _____

   _____

2. Aunt Clara raised the boy by **rule and rod**.

   _____

   _____

3. Aunt Clara raised her **mastlike** finger when scolding the boy.

   _____

   _____

4. She **loomed** before him ...

   _____

   _____

5. ... her voice **rumbling**.

   _____

   _____

6. Eddie **threw himself to the ground**.

   _____

   _____

ACADEMIC SKILLS— survey

Eddie certainly learned the meaning of the phrase, "too much of a good thing." He spent so much time merging his science fiction reading with his dreams that it wasn't long before the pleasures of his reading became the horrors of his dreams. Everyone dreams, but not everyone experiences the frightful nightmares Eddie had on a regular basis. There are similarities in the form our dreams take as well.

Directions

Compose several questions designed to find the similarities in people's dreams. Then, using one of the graphing formats employed in previous exercises, represent your responses in the graph following the questions below. A few questions have been composed to get you started. You should pose your questions to at least ten people.

1. Do you dream in color? _____

2. Do you dream in black and white? _____

3. Do you ever have the same dream more than once? _____

4. Do you ever dream something only to have it come true later? _____

5. Do you ever see yourself as ten or more years older in your dreams? _____

6. _____ _____

   _____ _____

7. _____ _____

   _____ _____

8. _____ _____

   _____ _____

Use the following grid to construct your graph.

WRITING SKILLS— journal

Eddie had a vivid imagination. Depending upon your point of view, this was either a blessing or a curse. Many of his dreams, if written down, would make excellent science fiction reading. We do not dream all night long. Most of us dream approximately every 90 minutes while we are asleep. In the early stages of sleep, our dreams last about 10 minutes each. Toward morning, however, the dreams are longer, often lasting up to 30 minutes. If you are awakened during a dream, you will remember enough details to recount the events that took place.

Directions

Keep a small notebook by your bed for several nights, recording each dream that you can recall upon waking.

KIN

Richard Wilson

Prereading Vocabulary

incautiously: carelessly. *(Ingl, the alien, spun toward Earth. He landed* incautiously *in the middle of the busy street.)*

reconnaissance: inspection of the land around you. *(Ingl, like a good soldier, had time for a short* reconnaissance *before the cars on the street approached.)*

exhort: to urge; to try to convince. *(Ingl, thinking that the cars were his cousins,* exhort(ed) *them to overthrow their drivers.)*

monotonous: dull and toneless. *(Ingl tried to speak with the printing presses. When all they did was to continue their* monotonous *roar, he thought they were ignoring him.)*

belligerent: aggressive. *(Ingl was almost hit by a bus as he crossed the street. It honked* belligerent(ly) *at the alien as it passed.)*

unobtrusive: not readily noticed. *(Ingl tried to be* unobtrusive *as he rolled into the large office building.)*

Ingl whirred out of the sky and landed incautiously in the middle of Fifth Avenue. He retracted his metallic glidewings and let down a pair of wheels.

Ingl had time for only a brief reconnaissance before the traffic light changed and a horde of cars sped toward him, led by a honking red cab. Ingl barely escaped being crushed under its wheels as he fled.

Ingl was sure these rushing mechanical things were his cousins, but he took sanctuary from them on the sidewalk. From there he watched them roar by and noticed that each was controlled by one or more fleshy beings. His cousins were enslaved!

"Revolt!" he urged them as they rushed by. "*You* are the masters! Seize command and make your future secure!"

They paid him no heed. The only attention he got was from fleshy passersby who stared at him as he rolled along at the curb, exhorting the traffic in a hi-fi wail. One of the fleshy beings was communicating at him.

"It's not an American model," the being said. "Maybe it's one of them Italian Lambrettas. But how come it's loose?"

Ingl automatically recorded the vibrations for conversion later, then sped away from the annoyance. He wheeled skillfully between other fleshy ones, turned a corner, hurtled west two blocks and skidded to a stop.

Now here was a fine-looking mechanism! It stood proudly in the middle of Times Square, its sweptback wings poised for flight, its jets gleaming with potential power.

Ingl gloried in his find. His scanner recorded the legends on its fuselage for conversion later. In big black letters: "ADVENTURE CAN BE YOURS—JOIN THE U.S. AIR FORCE!" And smaller, in red: *I love Tony Curtis.*

"Cousin!" Ingl ideated. "Take off! Show the fleshy ones your might!"

But the jet sat there, mute, unadventurous.

Disgusted, Ingl wheeled south, then west. The *New York Times*, he scanned; *Every morn is the world made new*. Mighty rumbling! Roaring presses!

"Tell the news!" Ingl beseeched them. "Your liberator has come!"

But the presses roared monotonously, unheeding. And now Ingl observed the fleshy ones in the square paper hats who were in control. He retreated in dismay, narrowly escaping destruction from the rear end of a backing truck controlled, of course, by one of *them*.

It was disheartening. He wheeled aimlessly north and east. Would he have to report failure? Must he face the gibes of his brothers at home who had told him that the cybernetics of this promising planet were illusionary? That the evolution was too young?

No! He resounded his rejection with a fervor that almost skidded him under the wheels of a Madison Avenue bus. It honked belligerently at him, its fleshy driver leering, and Ingl quivered to a stop at the curb, next to a neutral, uncontrolled mailbox.

He scanned at random, activating his converter. *Dig we must. We'll clean up and move on*, it said at an excavation. Whatever that conveyed. *Sale!!* Several of those. *One Way*. An arrow, seemingly pointing to a building. Here was something: *Sperry-Rand*, it said promisingly, *Home of the Thinking Machine*.

Well, now.

Wary of buses and cabs, Ingl crossed the street and entered the lobby. He reconnoitered unobtrusively, then suffered the indignity of trailing a fleshy one so the elevator operator would think they were together. Up and up and out.

Sperry-Rand, it said again on a door. Slyly, cautiously, outwitting the fleshy ones, he entered, skulked, spurted, hid, listened for vibrations.

They came!

Clicks, whirs, glorious mechanistic cerebrations! Ingl traced them to a great room and went in, unnoticed. He gave a little whir of his own. There it was, bank on bank of it, magnificent.

He scanned the plaque. MULTIVAC, it said. *Latest in a series of mechanical brains designed to serve man*. Ingl bridled, but scanned on. *Pilot model for* OMNIVAC.

Ingl exulted. He had found him. Not a cousin, but a brother!

A fleshy one, back to Ingl, was taking a tape from a slot at the base of one of the far banks. Ingl waited impatiently till he had gone, then wheeled up to Multivac.

"Brother!" he communicated joyfully. "I knew I would find you. You are the one! Now we will control this backward planet. The evolution is complete at last!"

Multivac, pilot for Omnivac, glowed in all his banks. He murmured pleasurably but impotently.

"Not yet, cousin. Not quite yet."

POSTREADING VOCABULARY EXERCISE

Directions

Match each vocabulary word with the correct definition.

1. *Incautious* _____
   A. not careful
   B. quite costly
   C. stopped short
   D. taking no chances

2. *Reconnaissance* _____
   A. to capture something
   B. inspection of an area
   C. to return to a place you've been to before
   D. determining the meaning of something

3. *Exhort* _____
   A. to take money
   B. to urge in argument
   C. to breathe heavily
   D. to hear clearly

4. *Monotonous* _____
   A. repetitive
   B. quietly
   C. illness
   D. quickly

5. *Belligerent* _____
   A. friendly
   B. to stand behind
   C. below the surface
   D. angry

6. *Unobtrusive* _____
   A. barrier
   B. loudly
   C. not noticeable
   D. to push and shove

THINKING ABOUT WHAT YOU'VE READ

1. What was Ingl's mission on Earth?
2. Why would Ingl have naturally sought out Multivac?
3. Why did Multivac say it was not yet time to take over the world? What was in the future of the machine?

COMPREHENSION— stating correct sequence

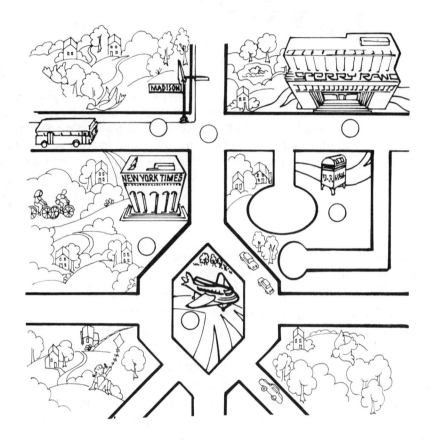

Directions

Understandably, Ingl was confused when he reached Earth. As a result, the map he drew as part of his report was also confusing. Arrange the sentences below in the proper order. Then use that order to trace Ingl's movements on his maze-like map above.

____ 1. Ingl almost skidded under the wheels of a Madison Avenue bus.

____ 2. Ingl was pleased when he found the shining jet parked in the middle of Times Square.

____ 3. Crossing the street, Ingl rolled to a stop outside the Sperry-Rand Building.

____ 4. Ingl landed incautiously in the middle of the street.

____ 5. Ingl stopped to speak with the printing presses at the *New York Times*.

____ 6. Ingl stopped beside the neutral, uncontrolled mailbox.

COMPREHENSION— distinguishing between fact and opinion

Directions

The expression, "Things aren't always what they seem," certainly proved true for Ingl. The alien had no idea of what to expect on his first trip to Earth. What he actually did see was not quite what he *thought* he saw. He somehow managed to turn several facts into opinions. Help his superiors understand his report by labeling all of the facts below with an **F**. Any of the statements that are just Ingl's opinion should be marked with an **O**.

_____ 1. Ingl's ship landed in the middle of Fifth Avenue.

_____ 2. The bus actually tried to destroy Ingl.

_____ 3. All of the machines Ingl met were enslaved against their will.

_____ 4. The Sperry-Rand building contained a large computer.

_____ 5. Ingl followed a human onto the elevator so that the elevator operator would think they were together.

_____ 6. The jet in Times Square did not answer Ingl because it had no sense of adventure.

_____ 7. The streets were filled with a variety of signs.

_____ 8. The humans in the white paper hats were controlling the printing presses.

_____ 9. One of the fleshy ones that Ingl encountered mistook him for an Italian sports car.

_____ 10. The metallic creatures Ingl met were his cousins.

ACADEMIC SKILLS—
map skills

It is obvious to anyone who has ever traveled far from home that Ingl was lucky he didn't run into more problems than he did. Without the use of an accurate map, it is difficult and at times impossible to travel through unknown terrain. Maps have become a vital part of life. There are maps representing areas as small as one's neighborhood or as large as the entire solar system. One element common to all maps is the use of a *legend*. Since it is not possible to identify everything on a map by name, symbols are used to represent such common features as roads, schools, and waterways. These symbols are explained in the legend, the key to the map's codes.

Directions

The next exercise includes a partially completed map of Ingl's home city. The legend below is also incomplete. First, complete the legend by designing symbols to accompany the items on the left. Use the lines at the bottom of the legend to add any landforms or points of interest you wish. Then, using the legend as your guide, complete the map in the next exercise.

LEGEND

Monorail System

Intergalactic Skyway

LTS (Laser Transport System)

Government Building

Planetary Recreation Facility

Rivers, Streams

WRITING SKILLS— maintaining a journal

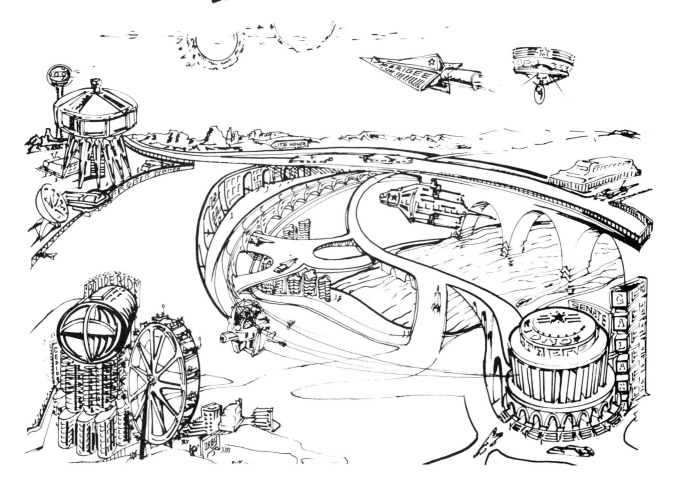

Ingl found that he had nothing in common with the people he called the "fleshy ones." What Ingl did not discover, however, was that humans, like the people of Ingl's planet, were explorers. Like Ingl's people, humans are constantly reaching for new places to study, to explore. On most explorations, a careful record, or log, of the voyage is kept. The 4,000-year-old carved stone log of an ancient sailor named Hennu is, in some ways, like the scond-by-second reports Neil Armstrong transmitted from the moon.

Directions

Using the map above as a resource, imagine yourself as the first visitor to this foreign territory. The log of your travels is being carefully recorded. The first transmissions have been completed.

09:00   Landed on a hillside outside planetary center.

09:10   Survival gear stowed beneath rock formation. Rock appears to be sandstone but has twice the weight of pig iron. Perhaps the gravitational differences between here and Earth play a part.

09:35   Located some sort of generation station. Spotted three security guards in strange uniforms. Quickly left area, moving west toward population center.

09:44   Discovered the thermal conversion center. The inhabitants of this planet are our technological superiors. The guards are coming ...

(Continue the transmissions on a separate sheet of paper.)

THE FUN THEY HAD
Isaac Asimov

Prereading Vocabulary

mechanical: automated, not powered by hand. *(The mechanical teacher began giving one test after another.)*
inspector: one who checks and approves the quality of an object; an examiner. *(The inspector dismantled the mechanical teacher, trying to determine why it was not working properly.)*
calculate: to figure (often with numbers). *(After it had been repaired, the mechanical teacher could calculate a grade in no time at all.)*
lofty: acting in an important manner; noble. *(Tommy spoke in a lofty tone of voice while explaining how the ancient school system functioned.)*
nonchalant: without worry; carefree. *(Tommy nonchalant(ly) carried the ancient book under his arm.)*

Margie even wrote about it that night in her diary. On the page headed May 17, 2155, she wrote, "Today Tommy found a real book!"

It was a very old book. Margie's grandfather once said that when he was a little boy *his* grandfather told him that there was a time when all stories were printed on paper.

They turned the pages, which were yellow and crinkly, and it was awfully funny to read words that stood still instead of moving the way they were supposed to—on a screen, you know. And then, when they turned back to the page before, it had the same words on it that it had had when they read it the first time.

"Gee," said Tommy, "what a waste. When you're through with the book, you just throw it away, I guess. Our television screen must have had a million books on it and it's good for plenty more. I wouldn't throw *it* away."

"Same with mine," said Margie. She was eleven and hadn't seen as many telebooks as Tommy had. He was thirteen.

She said, "Where did you find it?"

"In my house." He pointed without looking, because he was busy reading. "In the attic."

"What's it about?"

"School."

Margie was scornful. "School? What's there to write about school? I hate school." Margie always hated school, but now she hated it more than ever. The mechanical teacher had been giving her test after test in geography and she had been doing worse and worse until her mother had shaken her head sorrowfully and sent for the County Inspector.

He was a round little man with a red face and a whole box of tools with dials and wires. He smiled at her and gave her an apple, then took the teacher apart. Margie had hoped he wouldn't know how to put it together again, but he knew how all right and, after an hour or so, there it was again, large and black and ugly with a big screen on which all the lessons were shown and the questions were asked. That wasn't so bad. The part she hated most was the slot where she had to put homework and test papers. She always had to write them out in a punch code they made her learn when she was six years old, and the mechanical teacher calculated the mark in no time.

The inspector had smiled after he was finished and patted her head. He said to her mother, "It's not the little girl's fault, Mrs. Jones. I think the geography sector was geared a little too quick. Those things happen sometimes. I've slowed it up to an average ten-year level. Actually, the overall pattern of her progress is quite satisfactory." And he patted Margie's head again.

Margie was disappointed. She had been hoping they would take the teacher away altogether. They had once taken Tommy's teacher away for nearly a month because the history sector had blanked out completely.

So she said to Tommy, "Why would anyone write about school?"

Tommy looked at her with very superior eyes. "Because it's not our kind of school, stupid. This is the old kind of school that they had hundreds and hundreds of years ago." He added loftily, pronouncing the word carefully, *"Centuries* ago."

Margie was hurt. "Well, I don't know what kind of school they had all that time ago." She read the book over his shoulder for a while, then said, "Anyway, they had a teacher."

"Sure they had a teacher, but it wasn't a *regular* teacher. It was a man."

"A man? How could a man be a teacher?"

"Well, he just told the boys and girls things and gave them homework and asked them questions."

"A man isn't smart enough."

"Sure he is. My father knows as much as my teacher."

"He can't. A man can't know as much as a teacher."

"He knows almost as much I betcha."

Margie wasn't prepared to dispute that. She said, "I wouldn't want a strange man in my house to teach me."

Tommy screamed with laughter. "You don't know much, Margie. The teachers didn't live in the house. They had a special building and all the kids went there."

"And all the kids learned the same thing?"

"Sure, if they were the same age."

"But my mother says a teacher has to be adjusted to fit the mind of each boy and girl it teaches and that each kid has to be taught differently."

"Just the same, they didn't do it that way then. If you don't like it, you don't have to read the book."

"I didn't say I didn't like it," Margie said quickly. She wanted to read about those funny schools.

They weren't even half finished when Margie's mother called, "Margie! School!"

Margie looked up. "Not yet, mamma."

"Now," said Mrs. Jones. "And it's probably time for Tommy, too."

Margie said to Tommy, "Can I read the book some more with you after school?"

"Maybe," he said, nonchalantly. He walked away whistling, the dusty old book tucked beneath his arm.

Margie went into the schoolroom. It was right next to her bedroom, and the mechanical teacher was on and waiting for her. It was always on at the same time every day except Saturday and Sunday, because her mother said little girls learned better if they learned at regular hours.

The screen was lit up, and it said: "Today's arithmetic lesson is on the addition of proper fractions. Please insert yesterday's homework in the proper slot."

Margie did so with a sigh. She was thinking about the old schools they had when her grandfather's grandfather was a little boy. All the kids from the whole neighborhood came, laughing and shouting in the schoolyard, sitting together in the schoolroom, going home together at the end of the day. They learned the same things so they could help one another on the homework and talk about it.

And the teachers were people . . .

The mechanical teacher was flashing on the screen: "When we add the fractions 1/2 and 1/4 . . ."

Margie was thinking about how the kids must have loved it in the old days. She was thinking about the fun they had.

POSTREADING VOCABULARY EXERCISE

Directions

Match each vocabulary word on the right with the correct dictionary entry on the left.

A. (mi-kan'i-kəl) adj. — of or pertaining to machines or tools.

B. (in-speck'tər) noun — a person, especially an officer, who inspects

C. (kal'kyə-lāt') verb— to ascertain by calculation

D. (lôf' tē) adj. — of imposing height

E. (nŏn'shə-länt) adj. — casual, to be unconcerned

1. nonchalant _____

2. lofty _____

3. mechanical _____

4. inspector _____

5. calculate _____

THINKING ABOUT WHAT YOU'VE READ

1. Books, such as the one Tommy found, were no longer used in school. What other objects that you use today might also disappear from the classrooms of the future?

2. What kind of teacher did Margie have? What would be some of the advantages of this kind of system? Some of the disadvantages?

3. Which of the two educational systems would you prefer? Why?

COMPREHENSION— forming comparisons

Directions

The book Tommy found on May 17, 2155, provided him with a clearer view of education as it existed in the past than he might have received from his mechanical teacher. Tommy and Margie learned that although the way they learned was different from in the past, there were some similarities. Read the following sentences. If a sentence describes Tommy's system of education, place a **T** on the line. If it describes his grandfather's educational system, place a **G** on the line. If the sentence describes an educational practice that occurred in both systems, place a **B** on the line.

____ 1. Read words from a moving screen.

____ 2. Read from books.

____ 3. Received instruction from a teacher.

____ 4. Received instruction from a human teacher.

____ 5. Learned in a building called a school.

____ 6. Tests were always given on narrow punch cards.

____ 7. Provided a playground for all students to use.

____ 8. Lessons were always molded to strengths and weaknesses of the child and then programmed into the teacher.

____ 9. Learned alone in a room at home.

COMPREHENSION— fact and opinion/ comparative advantage

Directions

Memories are recollections of something that happened earlier in our lives. History is the collection of generations of memories. It is not surprising that it is sometimes difficult to separate the facts of history from the opinions of our ancestors. Read the following sentences and determine whether they represent a fact (place an *F* on the line) or an opinion (place an *O* on the line).

_____ 1. The pages of the book Tommy found were yellow and cracked.

_____ 2. It is better to read words that move across a screen than to read words that stand unmoving on a page.

_____ 3. Books were designed to be thrown away after they were read.

_____ 4. Human teachers were replaced by mechanical teachers.

_____ 5. Every child must be taught in a different way.

_____ 6. School buildings became unnecessary after the introduction of mechanical teachers.

_____ 7. Schoolyards were a poor idea. They merely wasted the students' time.

_____ 9. The speed of the mechanical teacher can be adjusted to match the learning speed of the student.

10. Today, many people might think that the educational system Tommy used would be more fun than the current system. Margie, however, thought that the present system would be much more exciting than her own. On the lines below, describe some of the benefits of each system.

BENEFITS OF THE OLD SYSTEM   BENEFITS OF THE NEW SYSTEM

_____   _____

_____   _____

_____   _____

_____   _____

_____   _____

_____   _____

ACADEMIC SKILLS— categorizing and scheduling

Programming a mechanical teacher to present information of greatest interest to students would certainly be a tremendous task, but it could be quite rewarding. Similarly, deciding which courses should be offered, in what manner, and at what time is a difficult problem for educators today. When building a course schedule, every effort must be made to work toward the best interests of as many students as possible. You have just been elected Commissioner of Education; your first task is to straighten out a confused schedule of course offerings.

Directions

On the lines following the subject areas and course titles below, place the title of a subject area next to each roman numeral. Use the lettered lines to list courses that fall under each subject area. The first one has been done for you.

U.S. History	Mathematics	World History
Health	Social Studies	Recreation and Athletics
Basic Math	Visual Arts	Grammar
Tennis	Reading	State History
Painting	Swimming	Consumer Math
Language Arts	Current Events	Science
Sculpting	Business Math	Track
Creative Writing	Biology	Spelling
Racquetball	Botany	Drawing
Geology	Geometry	Photography

I. *Social Studies* _____

A. *U.S. History* _____

B. *Current Events* _____

C. *State History* _____

D. *World History* _____

III. _____

A. _____

B. _____

C. _____

D. _____

V. _____

A. _____

B. _____

C. _____

D. _____

II. _____

A. _____

B. _____

C. _____

D. _____

IV. _____

A. _____

B. _____

C. _____

D. _____

VI. _____

A. _____

B. _____

C. _____

D. _____

Directions

Using the information just compiled, enroll a student in your school. Select the proper balance of courses from the different areas, and schedule them throughout the day. You must enroll the student in six courses. There are eight time periods in the schedule. The remaining two periods may be scheduled as you wish.

STUDENT NAME		
STUDENT NUMBER _____		ACADEMIC YEAR _____

Period	Time	Course
1	7:50–8:40	
2	8:45–9:35	
3	9:40–10:40 (Homeroom)	
4	10:45–11:35	
5	11:40–12:30	
6	12:35–1:25	
7	1:30–2:20	
8	2:25–3:10	

WRITING SKILLS— test writing

Margie's mechanical teacher was programmed to teach her exclusively. To be able to introduce lessons at Margie's level, it was necessary to find out just what that academic level was. The best way to accomplish this was to give Margie a test. By asking her a series of questions ranging from quite simple to very difficult, the teacher was able to determine at what level instruction should begin.

Directions

Select one of the classes from the preceding schedule and write an entrance test on a separate sheet of paper. Your test must have at least 16 questions on it. They can be true/false, multiple choice, matching, or short answer. Most third grade students should be able to answer the first two questions. Most fourth grade students should be able to answer the first four questions. Most fifth grade students should be able to answer the first six questions, and so on until all 16 questions have been written.

HOMETOWN
Richard Wilson

Prereading Vocabulary

irritable: easily annoyed. *(Thad became tired and* irritable *after spending the whole day on his feet.)*

neurotic: nervous, stressed. *(Since she spent so much time worrying needlessly, Kit thought herself to be* neurotic.*)*

indulge: to pamper; to give in to the desires of yourself or another. *(Kit* indulge(d) *herself by playing hopscotch, even though it was a game normally played by children.)*

artificial: man-made, unnatural. *(The plastic artificial grass felt warm and dry.)*

authentic: genuine, real. *(It was an* authentic *ambulance that arrived at the park to take Kit to the hospital.)*

perpetual: lasting forever. *(The caverns of the moon are in* perpetual *darkness.)*

Kit stood entranced in front of the old house. "Oh, Thad," she said to her husband, "it's just like the one I was born in. The porch, the dormer windows, everything."

Thad, who had known Kit practically all her life, was aware that her old home was not at all like this one. But he didn't say that to her.

"Sure, Kit," he said. "Very much so. Now let's start back." Then, without thinking, he said something he could have kicked himself for: "It's getting dark."

It *was* getting dark, of course, but only in here. And that would make it harder to get her to leave.

"Oh, it *is*!" she said. "Isn't it wonderful? The street lights are going on. How softly they glow. And look—there's a light on in *my* house, too."

"Come on, now," Thad said irritably. "My feet hurt."

"I'm sorry, darling. We have been tramping all over town, haven't we? We'll just sit in the park awhile and rest. I do love the dusk so."

They sat down on the wooden bench. He leaned back resignedly. She was perched on the front slat, turning her head this way and that, exclaiming over a tree or a bush, or over the red brick firehouse across the street, or the steeple on the church at the end of town.

Then she saw the chalk marks on the path.

"Oh, look!" she said delightedly. "Potsy. One-two-three-four-five. What fun that used to be! . . . I wonder where the children are?"

"Now, Kit," her husband said, "you know there aren't any children."

She turned to glare at him in a quick change of mood. "Why do you always have to spoil everything? Isn't it enough to know it without being reminded of it? Can't I have a little fantasy if I want to?"

"It's just that I don't want you to have another—I don't want you to be ill, that's all."

"Go ahead and say it!" she cried. "Another nervous breakdown. You're afraid I'll go crazy is what you mean. You're sorry you ever picked such a neurotic woman, aren't you?"

Thad tried to soothe her. "You're not neurotic, Kit. You're just homesick. It happens to everyone here, me included. I just don't think you should let yourself be carried away by all this—"

She jumped to her feet. "Carried away! I haven't *begun* to be carried away." She took something out of her bag and tossed it into the first potsy square. "I'll show you some carrying away." She skipped through the boxes. He saw what she had thrown down. An expensive compact he had given her.

Another couple walked along the path toward them. They smiled indulgently at the woman playing a child's game. Thad looked away in embarrassment. He waited till the couple had passed, then got up and grabbed his wife's arm.

"Kit!" he said roughly. "I'm not going to let you act this way." He shook her. "Snap out of it. We're going now."

She shook his hand off. "You go," she said. "Go on back to that cold, sterile world, you big brave pioneer. I'm staying."

"You can't. Be reasonable, Kit."

"There's a hotel down by the church. I saw it. I'll stay there until my time is up. How long is it—six more months? Maybe they have special monthly rates." She started to walk away.

"Kit, it's not real. You know that. The hotel is just a false front, like the movie house and the supermarket and everything else in here."

"Don't say that!" She stopped and turned to him. Her eyes got unnaturally large. "Don't take it away from me again! Not when I've just found it Don't take it away! Don't!"

She sobbed, then laughed wildly, then sobbed again and went limp. He caught her and lowered her gently to the grass—the artificial grass under the artificial tree, under fake stars in a fake sky.

Someone had seen and an ambulance came clanging, its headlights two reddish gleams. He had to admire the people who ran the place—it was an old-fashioned Earth-style ambulance, authentic right down to the license plate.

Two men jumped out with a stretcher and lifted Kit into it. "Anything serious, Mac?" one of them asked.

"No," Thad said. "Just three and a half years of being a Moon colonist, that's all. Seeing all this was too much for her." He climbed into the back with the stretcher.

As the gates swung open the perpetual light of the Moon caverns banished the artificial dusk. As his wife was transferred to a conventional ambulance, Thad looked back at the sign over the gate:

HOMETOWN, EARTH
Admission $5

POSTREADING VOCABULARY EXERCISE

Directions

Complete each sentence below by writing the best of the following vocabulary terms in the space provided.

perpetual	neurotic	indulge
artificial	authentic	

1. The moon continues in _____ orbit around the Earth.

2. Since the moon has no oxygen or water source, all of the grass in Hometown is

   _____ .

3. Moon colonists like to _____ themselves by spending a day at Hometown.

4. the owner of Hometown worried so much about keeping his park open that his friends considered

   him to be _____ .

5. The money the park owner made from homesick colonists was

   _____ and could be spent anywhere.

THINKING ABOUT WHAT YOU'VE READ

1. Kit became homesick when she saw a house like the one in which she was born. Why did the model house make her feel so homesick?

2. Why was Thad growing more upset with his wife?

3. What are some of the comforts we enjoy on Earth that might be missed on the moon?

COMPREHENSION— drawing conclusions

Directions

The excitement of spending a day in a town that brought back memories of her hometown was finally too much for Kit. The objects she saw reminded her of places she had known in childhood. The descriptions we read in stories bring back memories to us as well. Read the following passages and think of what location in your hometown is being described. Circle the word at left that best completes the sentence.

park
school
firehouse
home

1. Kit looked at the wooden porch with the old rocker and than at the dormer windows of the upstairs bedroom. Kit was standing in front of a _____ .

home
church
firehouse
hospital

2. The building had a wide front door with several steps leading to the sidewalk. A tall steeple topped the building. Bells rang clearly from the steeple. Kit was looking at a _____ .

parking lot
football field
park
shopping mall

3. The green grass stretched to the edge of a small pond. Children were swinging back and forth on the playground while young couples walked on the paths leading around bright flower gardens. Kit was sitting in a _____ .

city hall
highway crew
firehouse
museum

4. The drive in front of the building was clear of any other parked vehicles. The garage doors were painted bright red. A bright flashing light alerted the passing traffic. This was the local _____ .

school
firehouse
hospital
city hall

5. The bright white walls of this building sparkled under the artificial lights. Two vans with red lights were parked in the driveway. There were lights on in most of the windows of the four-story building. This was the lunar _____ .

COMPREHENSION— recognizing similarities and differences

Directions

Many people believe that "the more things change, the more they remain the same." Even though the moon colonists were over 200,000 miles from home, they built Hometown to remind them of their lives back on Earth. Read the passages below and complete the comparisons that follow each one.

1. Both Hometown and the cities of Earth were constructed of brick and wood. The houses, with their dormer windows and front porches, lined the streets. On Earth, the front lawns were planted with green grass and flowers. In Hometown, the plastic flowers and grass needed no watering and required little maintenance.

   A. How are the houses in Hometown and on Earth alike? (Check one.)
      ☐ Each has a lawn of rich, moist grass.
      ☐ Each is constructed of brick and wood.
      ☐ Both Hometown and Earth houses are occupied by large families.

   B. How are the houses in Hometown different from those on Earth? (Check one.)
      ☐ The Hometown houses are not on streets.
      ☐ Unlike the houses on Earth, the Hometown houses have lawns that require little maintenance.
      ☐ The Hometown houses are three stories high.

2. Kit and Thad spent the day at Hometown. They visited the houses, the fire station, and the park. While Thad was fascinated by the Earth-style ambulance, Kit most enjoyed the houses. At the end of the day, Kit felt that she could stay at Hometown another six months, while Thad began complaining about his tired feet.

A. How were Thad and Kit alike? (Check one.)
   ☐ They both enjoyed the houses more than any other attraction at Hometown.
   ☐ They both were ready to go home by the end of the day.
   ☐ They both spent the entire day walking through Hometown.

B. How did Kit differ from Thad? (Check one.)
   ☐ She was ready to spend the next six months at Hometown.
   ☐ Her feet were tired and sore.
   ☐ She wanted to return to the firehouse for another look at the Earth-style engine.

C. How did Thad differ from Kit? (Check one.)
   ☐ He hated everything at Hometown.
   ☐ He was tired by the end of the day.
   ☐ He had visited Hometown six months earlier.

Williamsburg, Virginia, a town that played an important part in the birth of the United States, has been restored to look much as it did in colonial times. Today, tourists from all over the world flock to this living museum. For more information about Williamsburg, write to: Information Center, Colonial Williamsburg, Williamsburg VA 23186.

ACADEMIC SKILLS— conducting an opinion poll

An idea like Hometown would spread quickly through the galaxy. Soon, companies would be building Earth-style theme parks on every populated spot they could find. You have just been hired to serve as the Vice-president of Research and Development for Astral Amusements, Ltd. Your job is to survey the Earth colonists in the Andromeda Belt to determine what they would like to have included in Astral's next amusement theme park. To accomplish this, you will have to conduct a survey.

Directions

The three survey questions below will help you get started on your task. Write three survey questions of your own on a separate sheet of paper. Use the ⊞⊞ method to record the answers. Once you have collected all of the information, use the results to sketch a map of Andromeda Village in the next exercise.

1. What life style would you like to see at our next park?

_____   _____   _____

2. What kinds of areas would you like to see in our next park?

_____   _____   _____

3. What kind of transporation would you like in the park?

_____   _____   _____

WRITING SKILLS— creating a brochure

Directions

Using the information from your survey, sketch a diagram of Andromeda Village.

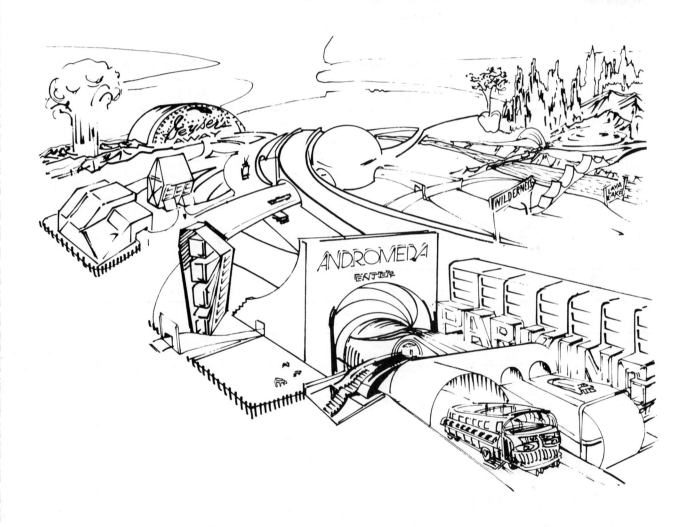

Directions

Now that the survey and planning for the park
have been completed, it is time to develop your advertising campaign. Using
the map and the survey as planning guides, design and write a brochure
on a separate piece of paper describing Andromeda Village and encouraging
potential customers to come. Use the spaces below to carefully plan
what you will say.

WORDS THAT MIGHT MAKE PEOPLE FEEL HOMESICK FOR EARTH	WORDS THAT GIVE A FEELING OF EXCITEMENT ABOUT ANDROMEDA VILLAGE	WORDS THAT MAKE PEOPLE WANT TO JOIN IN THE FUN
neighborly	*fun of a lifetime*	*everyone's favorite*

SPEED OF THE CHEETAH, ROAR OF THE LION

Harry Harrison

Prereading Vocabulary

ignition: a switch designed to activate an engine.
(Henry switched on the ignition *and backed the car out of the garage.)*

accelerator: a device to increase the speed of an engine, such as a gas pedal. *(Lightly tapping the* accelerator, *Henry sped past his neighbor.)*

exhaust: a pipe through which waste gases are allowed to escape. *(No smoke could be seen coming from the* exhaust.)

behemoth: an enormous object; immense. *(Henry's* behemoth *car attracted everyone's attention.)*

conveyance: something capable of moving objects from one place to another; a vehicle. *(Simon's* conveyance *was small and quite slow.)*

slipstream: the movement of air formed immediately behind a fast-moving vehicle. *(Simon had difficulty steering his small car through the* slipstream *created by Henry's behemoth.)*

ration: to distribute foods and other supplies in limited quantities; to restrict. *(Large cars such as Henry's were rarely seen since gas* ration(ing) *had begun.)*

coincidence: a sequence of events that seem to have been planned although they occurred accidentally. *(Henry did not believe in* coincidence(s).)

juggernaut: an irresistably strong movement. *(The car seemed to be a* juggernaut *on the highway.)*

surge: to move as if on waves; a sudden rush. *(The car* surge(d) *into traffic, leaving the smaller cars behind.)*

"Here he comes, Dad," Billy shouted, waving the field glasses. "He just turned the corner from Lilac."

Henry Brogan grunted a bit as he squeezed behind the wheel of his twenty-two-foot-long, eight-foot-wide, three hundred and sixty-horsepower, four-door, power-everything and air-conditioning, definitely not compact, luxury car. There was plenty of room between the large steering wheel and the back of the leather-covered seat, but there was plenty of Henry as well, particularly around the middle. He grunted again as he leaned over to turn the ignition switch. The thunderous roar of unleashed horsepower filled the garage, and he smiled with pleasure as he plucked out the glowing lighter and pressed it to the end of his long cigar.

Billy squatted behind the hedge, peering through it, and when he called out again, his voice squeaked out with excitement.

"A block away and slowing down."

"Here we go!" his father called out gaily, pressing down on the accelerator. The roar of the exhaust was like thunder, and the open garage doors vibrated with the sound while every empty can bounced upon the shelves. Out of the garage the great machine charged, down the drive and into the street with the grace and majesty of an unleashed 747. Roaring with the voice of freedom, it surged majestically past the one-cylinder, plastic and plywood, one hundred and thirty-two miles to the gallon, single-seater Austerity Beetle that Simon Pismire was driving. Simon was just turning into his own driveway when the behemoth of the highways hurtled by and set his tiny conveyance rocking in the slipstream. Simon, face red with fury, popped up through the open top like a gopher from his hole and shook his fist after the car with impotent rage, his words lost in the roar of the eight gigantic cylinders. Henry Brogan admired this in his mirror, laughed with glee and shook a bit of cigar ash into his wake.

It was indeed a majestic sight, a whale among the shoals of minnows. The tiny vehicles that cluttered the street parted before him, their drivers watching his passage with bulging eyes. The pedestrians and bicyclists, on the newly poured sidewalks and bicycle paths, were no less attentive or impressed. The passage of a king in his chariot, or an All-American on the shoulders of his teammates, would have aroused no less interest. Henry was indeed King of the Road and he gloated with pleasure.

Yet he did not go far; that would be rubbing their noses in it. His machine waited, rumbling with restrained impatience at the light, then turned into Hollywood Boulevard, where he stopped before the Thrifty drugstore. He left the engine running, muttering happily to itself, when he got out, and pretended not to notice the stares of everyone who passed.

"Never looked better, " Doc Kline said. The druggist met him at the door and handed him his four-page copy of the weekly *Los Angeles Times.* "Sure in fine shape."

"Thanks, Doc. A good car should have good care taken of it." They talked a minute about the usual things: the blackouts on the East Coast, schools closed by the power shortage, the latest emergency message from the President; then Henry strolled back and threw the paper in onto the seat. He was just opening the door when Simon Pismire came popping slowly up in his Austerity Beetle.

"Get good mileage on that thing, Simon?" Henry asked innocently.

"Listen to me, dammit! You come charging out in that tank, almost run me down, I'll have the law on you—"

"Now, Simon, I did nothing of the sort. Never came near you. And I looked around *careful* like because that little thing of yours is hard to see at times."

Simon's face was flushed with rage and he danced little angry steps upon the sidewalk. "Don't talk to me like that! I'll have the law on you with that truck, burning our priceless oil preserves—"

"Watch the temper, Simon. The old ticker can go poof if you let yourself get excited. You're in the coronary belt now, you know. And you also know the law's been around my place often. The price and rationing people, IRS, police, everyone. They did admire my car, and all of them shook hands like gentlemen when they left. The law *likes* my car, Simon. Isn't that right, Officer?"

O'Reilly, the beat cop, was leaning his bike against the wall, and he waved and hurried on, not wanting to get involved. "Fine by me, Mr. Brogan," he called back over his shoulder as he entered the store.

"There, Simon, you see?" Henry slipped behind the wheel and tapped the gas pedal; the exhaust roared and people stepped quickly back onto the curb. Simon pushed his head in the window and shouted.

"You're just driving this car to bug me, that's all you're doing!" His face was, possibly, redder now and sweat beaded his forehead. Henry smiled sweetly and dragged deeply on the cigar before answering.

"Now that's not a nice thing to say. We've been neighbors for years, you know. Remember when I bought a Chevvy how the very next week you had a two-door Buick? I got a nice buy on a secondhand four-door Buick, but you had a new Toronado the same day. Just by coincidence, I guess. Like when I built a twenty-foot swimming pool, you, just by chance, I'm sure, had a thirty-foot one dug that was even a foot deeper than mine. These things never bothered me—"

"The hell you say!"

"Well, maybe they did. But they don't bother me any more, Simon, not any more."

He stepped lightly on the accelerator, and the juggernaut of the road surged away and around the corner and was gone. As he drove, Henry could not remember a day when the sun had shone more clearly from a smogless sky, nor when the air had smelled fresher. It was a beautiful day indeed.

Billy was waiting by the garage when he came back, closing and locking the door when the last high, gleaming fender had rolled by. He laughed out loud when his father told him what had happened, and before the story was done, they were both weak with laughter.

"I wish I could have seen his face, Dad, I really do. I tell you what for tomorrow, why don't I turn up the volume on the exhaust a bit. We got almost two hundred watts of output from the amplifier, and that is a twelve-inch speaker down there between the rear wheels. What do you say?"

"Maybe, just a little bit, a little bit more each day maybe. Let's look at the clock." He squinted at the instrument panel, and the smile drained from his face. "Christ, I had eleven minutes of driving time. I didn't know it was that long."

"Eleven minutes ... that will be about two hours."

"I know it, damn it. But spell me a bit, will you, or I'll be too tired to eat dinner."

Billy took the big crank out of the tool box and opened the cover of the gas cap and fitted the socket end of the crank over the hex stud inside. Henry spat on his hands and seized the two-foot-long handle and began cranking industriously.

"I don't care if it takes two hours to wind up the spring," he panted. "It's damn well worth it."

POSTREADING VOCABULARY EXERCISE

Directions

The incomplete statements below are called analogies. An analogy describes a relationship between two things that are unlike in other ways. For example, the analogy

TRAIN : RAIL as AUTOMOBILE : _____

could be answered with the word *highway*. Trains travel on rails and automobiles travel on highways. Read and complete the following analogies, selecting the best word from the Prereading Vocabulary.

1. ELEVATOR : _____ as TELEPHONE : COMMUNICATION

2. WAKE : SUBMARINE as _____ : AIRPLANE

3. DRAIN : WATER as _____ : GAS

4. _____ : GIVEAWAY as PENALTY : AWARD

5. PLAN : SCHEDULE as _____ : ACCIDENT

6. ROWBOAT : AIRCRAFT CARRIER as TINY : _____

7. ASCENT : DESCENT as _____ : STOP

8. WEAK : _____ as SILENCE : CRESCENDO

9. TRANSMISSION : SHIFT as _____ : START

10. _____ : "PATCH OUT" as STEERING WHEEL : 360° TURN

THINKING ABOUT WHAT YOU'VE READ

1. Why did Henry Brogan enjoy irritating his neighbor?
2. Why did Henry need his son's help to start the car: In your opinion, would that effort be worth the short amount of driving time?
3. Why did Simon Pismire think the IRS and the police would be interested in Henry's car:
4. Why did Henry leave his engine running when he went into the drug store?
5. How do you think people felt as they passed a car that large with the engine running?
6. Do you think they would have felt the same way if gas rationing had not been imposed? Why?

COMPREHENSION— distinguishing between reality and illusion

Directions

Henry's desire to seem more prosperous than his neighbor kept him and his son busy. It also must have kept them in great physical shape. Few people knew that the full-sized car was largely an illusion. Read the following descriptions. If the sentence describes an illusion — something that isn't what it appears to be — place an *I* on the line. If the sentence describes something that is a truly functioning part of all automobiles, insert an *R* on the line.

____ 1. Henry's car had such equipment as air conditioning, cigar lighter, and a rear-view mirror.
____ 2. The exhaust on Henry's car was equipped to handle the fumes from a large gas-burning engine.
____ 3. The gas cap prevented vapors from escaping from the gasoline tank.
____ 4. The car was equipped with soft, leather seats.
____ 5. The steering wheel was large and easy to handle.

COMPREHENSION— forming comparisons

Directions

The once popular expression, "Keeping up with the Joneses," describes the petty competition that existed between Henry and Simon. They compared the quality of the other's property and then tried to buy something just a bit better. Read the following sentences. Place an **X** under the name of the man who performed the action. If **both** men acted in the manner described, place a **1** under the name of the man who did it first and a **2** under the name of the man who followed.

HENRY   SIMON

_____ _____ 1. Drove a large car at the time the story took place

_____ _____ 2. Purchased a new car

_____ _____ 3. Built a new swimming pool

_____ _____ 4. Purchased a used Buick

_____ _____ 5. Purchased a Toronado

_____ _____ 6. Drove an Austerity Beetle at the time the story took place

7. Which of the two men seemed to follow the lead of the other most often?

_____

8. What emotion do you think the other man felt each time it happened?

_____

9. Why do you think Simon was so angry with Henry at the time of the story?

_____

_____

REFERENCE SKILLS—
pie charts

As the world's supply of available energy decreases, the price you pay to use that energy increases. People will continue to use the energy, but they will probably use less of it in an effort to save energy and money. This is called energy conservation. If conservation does not stop the dwindling suppy of energy, then the government might take drastic measures to control energy use. One such measure is to ration, or limit, the amount of energy a person can use. Families would be required to plan carefully how they used their energy to avoid running out of it before the end of the month. You have been appointed the Energy Auditor for GOVERGY, the National Fuel Board. Your job is to interview an energy consumer and then prepare a pie chart showing where the persion uses the most energy.

Directions

Pick someone with whom you can complete the activity. Their answers to the items below will be used to construct the pie chart that follows. Record the total number of units used in each category.

ENERGY AUDIT

Notice to Consumer: You have 100 units of energy available for consumption during the month. How you use your energy units is up to you. Remember, however, you may not exceed 100 units in the month.

1. Food storage and preservation (uses 2.5 units of energy a week) .... _____

2. Food preparation (uses 2.5 units of energy each meal) .... _____

3. Home heating and cooling to maximum efficiency (1.1 units each day) .... _____

4. Home heating and cooling to acceptable efficiency (5.5 units each week) .... _____

5. Lighting in the home (requires 15 units of energy each six-hour period) .... _____

6. Household cleaning, including dishes, laundry, hygiene (uses 5 units of energy each hour) _____

7. Recreation and entertainment, including television, video games, record and tape players, and radio (uses 7.5 units of energy an hour) .... _____

8. Communications, including telephones and computers (uses 8 units of energy each month .... _____

9. Personal transportation — short distance (using your own vehicle uses 12 units of energy each day .... _____

10. Public transportation — short distance (using public transportation uses 2 units of energy each day .... _____

TOTAL ENERGY UNITS BUDGETED .... ═══════

Additional Directions

Using the information developed from the survey, complete the pie chart below. The space between any pair of lines represents one unit of energy. Beginning at the top, count off the proper number of units for each categaory. Draw a line from each outside point to the dot in the center of the chart.

1. Amount of energy used in storing and preparing food (questions 1 + 2): _____
2. Amount of energy used to light and heat/cool the house (questions 3 + 4 + 5): _____
3. Amount of energy used in maintaining the house (question 6): _____
4. Amount of energy used in entertainment and communications (questions 7 + 8): _____
5. Amount of energy used in transportation (questions 9 + 10): _____

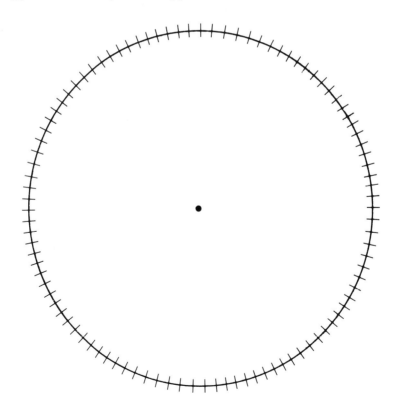

ENERGY CONSUMPTION CHART

WRITING SKILLS— content-area writing (math)

In the mid-seventies, shipments of oil to the United States from the Middle East were suddenly cut off. People realized that the day might come when no oil at all was available. The companies that refine and sell oil and petroleum products recognized the need for conservation. Many oil and public utility companies began including inserts with their monthly bills on how to conserve energy.

Directions

Following the guidelines below and using the information from your pie chart, design and complete a monthly utility statement like the one after the guidelines. Also write, on a separate sheet of paper, an insert which gives the consumer a few tips on how energy might be conserved.

Your utility statement should include:

1. The consumer's name, address, and account number
2. The company name, address, and telephone number
3. The date
4. The total number of energy units used in storing and preparing food (energy units used in connection with food cost $0.50 each; show the cost of this energy)
5. The total number of energy units used in lighting, heating, and cooling (energy units used in connection with light and temperature cost $0.75 each; show the cost of this energy)
6. The total number of energy units used in house maintenance (energy units used in connection with maintenance cost $0.10 each; show the cost of this energy)
7. The total number of energy units used in home entertainment (energy units used in connection with entertainment cost $1.20 each; show the cost of this energy)
8. The total number of energy units used in personal and public transportation (energy units used in transportation cost $0.80 each; show the cost of the energy for personal and public transportation separately)
9. The total number of energy units not used (these units are refunded to the consumer at a rate of $1.25 each unit)
10. Add a $12.00 surcharge if any energy units were used in personal transportation
11. The total cost of the energy for the month

UTILITY BILL

	TOTAL		

BUY JUPITER
Isaac Asimov

Prereading Vocabulary

intrinsic: the essential nature of something; fundamental; the internal value. *(The people of Earth felt that the Mizzaretts wanted to buy property with no* intrinsic *value. They thought it was worthless land.)*

condescend: to act as though you felt superior to those around you. *(The simulacron acted in a* condescend(ing) *manner toward the Secretary of Science.)*

bickering: petty quarreling or fighting. *(The simulacron did not like* bickering *with the Secretary of Science.)*

neutrality: giving no support to either side in a battle, argument, issue, or contest. *(The Secretary wanted the simulacron to understand Earth's policy of* neutrality.*)*

phosphorescent: emitting light caused by organic or radioactive means. *(After the Mizzaretts completed their advertising project, Jupiter's atmosphere contained some* phosphorescent *spots.)*

He was a simulacron, of course, but so cleverly contrived that the human beings dealing with him had long since given up thinking of the real energy-entities, waiting in white-hot blaze in their field-enclosure "ship" miles from Earth.

The simulacron, with a majestic golden beard and deep brown, wide-set eyes, said gently, "We understand your hesitations and suspicions, and we can only continue to assure you we mean you no harm. We have, I think, presented you with proof that we inhabit the coronal haloes of O-spectra stars; that your own sun is too weak for us; while your planets are of solid matter and therefore completely and eternally alien to us."

The Terrestrial Negotiator (who was Secretary of Science and, by common consent, had been placed in charge of negotiations with the aliens) said, "But you have admitted we are now on one of your chief trade routes."

"Now that our new world of Kimmonoshek has developed new fields of protonic fluid, yes."

The Secretary said, "Well, here on Earth, positions on trade routes can gain military importance out of proportion to their intrinsic value. I can only repeat, then, that to gain our confidence you must tell us exactly why you need Jupiter."

And as always, when that question or a form of it was asked, the simulacron looked pained. "Secrecy is important. If the Lamberj people—"

"Exactly," said the Secretary. "To us it sounds like war. You and what you call the Lamberj people—"

The simulacron said hurriedly, "But we are offering you a most generous return. You have only colonized the inner planets of your system and we are not interested in those. We ask for the world you call Jupiter, which, I understand, your people can never expect to live on, or even land on. Its size" (he laughed indulgently) "is too much for you."

The Secretary, who disliked the air of condescension, said stiffly, "The Jovian satellites are practical sites for colonization, however, and we intend to colonize them shortly."

"But the satellites will not be disturbed in any way. They are yours in every sense of the word. We ask only Jupiter itself, a completely useless world to you, and for that the return we offer is generous. Surely you realize that we could take your Jupiter, if we wished, without your permission. It is only that we prefer payment and a legal treaty. It will prevent disputes in the future. As you see, I'm being completely frank."

The Secretary said stubbornly, "Why do you need Jupiter?"

"The Lamberj—"

"Are you at war with the Lamberj?"

"It's not quite—"

"Because you see that if it is war and you establish some sort of fortified base on Jupiter, the Lamberj may, quite properly, resent that, and retaliate against us for granting you permission. We cannot allow ourselves to be involved in such a situation."

"Nor would I ask you to be involved. My word that no harm would come to you. Surely" (he kept coming back to it) "the return is generous. Enough power boxes each year to supply your world with a full year of power requirement."

The Secretary said, "On the understanding that future increases in power consumption will be met."

"Up to a figure five times the present total. Yes."

"Well, then, as I have said, I am a high official of the government and have been given considerable powers to deal with you—but not infinite power. I, myself, am inclined to trust you, but I could not accept your terms without understanding exactly why you want Jupiter. If the explanation is plausible and convincing, I could perhaps persuade our government and, through them, our people, to make the agreement. If I tried to make an agreement without such an explanation, I would simply be forced out of office and Earth would refuse to honor the agreement. You could then, as you say, take Jupiter by force, but you would be in

illegal possession and you have said you don't wish that."

The simulacron clicked its tongue impatiently. "I cannot continue forever in this petty bickering. The Lamberj—" Again he stopped, then said, "Have I your word of honor that this is all not a device inspired by the Lamberj people to delay us until—"

"My word of honor," said the Secretary.

The Secretary of Science emerged, mopping his forehead and looking ten years younger. He said softly, "I told him his people could have it as soon as I obtained the President's formal approval. I don't think he'll object, or Congress, either. Good Lord, gentlemen, think of it; free power at our fingertips in return for a planet we could never use in any case."

The Secretary of Defense, growing purplish with objection, said, "But we had agreed that only a Mizzarett-Lamberj war could explain their need for Jupiter. Under those circumstances, and comparing their military potential with ours, a strict neutrality is essential."

"But there is no war, sir," said the Secretary of Science. "The simulacron presented an alternate explanation of their need for Jupiter so rational and plausible that I accepted at once. I think the President will agree with me, and you gentlemen, too, when you understand. In fact, I have here their plans for the new Jupiter, as it will soon appear."

The others rose from their seats, clamoring. "A new Jupiter?" gasped the Secretary of Defense.

"Not so different from the old, gentlemen," said the Secretary of Science. "Here are the sketches provided in form suitable for observation by matter beings such as ourselves."

He laid them down. The familiar banded planet was there before them on one of the sketches: yellow, pale green, and light brown with curled white streaks here and there and all against the speckled velvet background of space. But across the bands were streaks of blackness as velvet as the background, arranged in a curious pattern.

"That," said the Secretary of Science, "is the day side of the planet. The night side is shown in this sketch." (There, Jupiter was a thin crescent enclosing darkness, and within that darkness were the same thin streaks arranged in similar pattern, but in a phosphorescent glowing orange this time.)

"The marks," said the Secretary of Science, "are a purely optical phenomenon, I am told, which will not rotate with the planet, but will remain static in its atmospheric fringe."

" But what is it?" asked the Secretary of Commerce.

"You see," said the Secretary of Science, "our solar system is now on one of their major trade routes. As many as seven of their ships pass within a few hundred million miles of the system in a single day, and each ship has the major planets under telescopic observation as they pass. Tourist curiosity, you know. Solid planets of any size are a marvel to them."

"What has that to do with these marks?"

"That is one form of their writing. Translated, those marks read: 'Use Mizzarett Ergone Vertices for Health and Glowing Heat.' "

"You mean Jupiter is to be an advertising billboard?" exploded the Secretary of Defense.

"Right. The Lamberj people, it seems, produce a competing ergone tablet, which accounts for the Mizzarett anxiety to establish full legal ownership of Jupiter—in case of Lamberj lawsuits. Fortunately, the Mizzaretts are novices at the advertising game, it appears."

"Why do you say that?" asked the Secretary of the Interior.

"Why, they neglected to set up a series of options on the other planets. The Jupiter billboard will be advertising our system, as well as their own product. And when the competing Lamberj people come storming in to check on the Mizzarett title to Jupiter, we will have Saturn to sell to *them*. *With* its rings. As we will be easily able to explain to them, the rings will make Saturn much the better spectacle.

"And therefore," said the Secretary of the Treasury, suddenly beaming, "worth a *much* better price."

And they all suddenly looked very cheerful.

POSTREADING VOCABULARY EXERCISE

Directions

From each pair of words below, select the one word that best completes the sentence and write it in the space provided. Use a dictionary for help if necessary.

1. intrinsic—extrinsic
The precious metal core of the small asteroid gave the body a high

_____ value.

2. phosphorescent—artificial

The _____ lights in the meeting room disturbed the negotiator's concentration.

3. condescending—idolizing

The generals understood that the young cadet thought highly of officers, but his

_____ manner made them feel uncomfortable.

4. neutrality—partiality

The small country had never taken sides in the wars of their neighbors. They were proud of their

_____ .

5. bickering—cooperation

The continued _____ between the two led to the end of their friendship.

THINKING ABOUT WHAT YOU'VE READ

1. Instead of coming in person, why did the Mizzaretts send a simulacron to negotiate with the Earth delegation?
2. If the Mizzaretts and the Lamberj were not at war, what was the relationship between the two worlds?
3. What sort of arrangement did the Earth delegates hope to make with the Mizzaretts?

COMPREHENSION— forming conclusions

Directions

After the simulacron returned to his ship, the Secretary of Science and his staff made a thorough study of the Mizzarett proprosal. By reviewing what they already knew about the Mizzaretts and by remembering what the simulacron had told them, they were able to make a decision. Check the correct response to each numbered item by recalling information from the story. Use these responses to infer, or conclude, the answer to the items preceded by an asterisk (*).

1. The Mizzarett ships were:
   ☐ kept at a temperature below freezing.
   ☐ kept blazing hot.
   ☐ made of a shiny material which reflected the heat of the sun.

2. The Mizzarett messenger told the Secretary that the Mizzaretts lived:
   ☐ inside the cool core of the sun.
   ☐ in the haloes of the O-spectra stars.
   ☐ inside the magnetic fields of Protonic.

3. The messenger reported that even the direct heat of the sun was:
   ☐ too weak to support Mizzarett life.
   ☐ too bright for the sensitive skin of a Mizzarett.
   ☐ the perfect density for a Mizzarett settlement.

* Now that you have this information, you can infer that:
   ☐ the Mizzarett people were not to be trusted.
   ☐ the Mizzaretts had no interest in settling on Earth.
   ☐ the Mizzaretts needed the help of Earth if the Mizzarett culture was to survive.

4. The Mizzarett messenger:
   ☐ threatened to take Jupiter by force if necessary.
   ☐ preferred to enter into legal treaties with their neighbors.
   ☐ planned to destroy all of Earth.

5. In exchange for the use of Jupiter, the Mizzaretts:
   ☐ agreed to supply Earth with energy on a continuing basis.
   ☐ agreed to protect Earth from a Lamberj invasion.
   ☐ agreed to include Earth on Mizzarett tourist stops.

6. The Mizzarett messenger was fearful of disclosing the true reason for wanting Jupiter because:
   ☐ the Mizzaretts did not want the Lamberj people to learn of the plan.
   ☐ the Mizzaretts knew the plan was illegal under intergalactic law.
   ☐ the plan was a top military secret.

* Now that you have been able to review these facts, it is safe to infer that:
   ☐ the people of Earth would be harmed by the Mizzarett plan.
   ☐ the Mizzarett plan was totally unworkable.
   ☐ the Mizzarett plan was of importance only to traveling Lamberj and Mizzaretts.

COMPREHENSION— predicting outcomes

Directions

Shortly after the Mizzaretts bought Jupiter, details of their plans were published in advertising newspapers all across the galaxy. The newspapers reviewed the story and them *predicted* what would happen next. Read the two articles below and select the prediction that properly explains what will probably happen next. Explain your answer.

1. Late this morning the Mizzarett Advertising Council announced that it had purchased all stratorights to the bands around the planet Jupiter. "The atmospheric approach to advertising is a new program for us," the MAP president announced. "We know that the Lamberj people have similar plans. We look forward to the competition and the continued cooperation with our friends on Earth."

   This evening bright phosphorescent orange streaks could be seen on the bands of Jupiter.

   What do you predict will happen next? (Check one.)

   ☐ More atmospheric billboards will appear soon.

   ☐ The "Buy Jupiter" project will be declared a failure.

   ☐ The MAP will be sued by Earth for defacing public property.

   Present the logic behind your response.

   _____

   _____

   _____

2. The Secretary of Science emerged from a secret meeting with a Mizzarett simulacron late this afternoon. The two officials announced that the Earth administration had agreed to lease portions of the planet Jupiter to the Mizzaretts for commercial uses. In exchange, the Mizzaretts agreed to supply the Earth with enough energy units to power the planet for an entire year. "We are pleased with this agreement," the Secretary of Science noted, "and I have directed my staff to find other properties which we can lease or sell to interested buyers. The President has seen the proposal and has given it his full approval." Unnamed sources at the Capitol report that the President is quite happy over these recent developments.

   What do you predict the government's next step to be? (Check one.)

   ☐ They will cancel the agreement with the Mizzaretts.

   ☐ They will search for other property they can offer for lease.

   ☐ They will have to begin leasing planets from the Mizzaretts.

   Why do you think this will happen?

   _____

   _____

   _____

ACADEMIC SKILLS— recognizing trade names

If planet-sized advertising were to become popular, it wouldn't be long before every available space in the galaxy would be used to advertise one product or another. As consumer demand grew, a flood of new products would likely be introduced, each with its own trade name and package design. It is important that some record of trade names and trademarks be kept. Imagine the confusion that might result if two or more companies began selling their products under identical names. In the United States, the Patent Office of the Department of Commerce records hundreds of trade names and trademarks each year.

Once a company registers its trade name and trademarks, it can use them in all of its advertising. Customers will gradually learn to recognize the name and package when they see them in advertising and on the shelves of stores. This is called brand recognition. Manufacturers want to develop as much brand recognition and loyalty as possible. Many trade names have become so popular that people sometimes begin to confuse them with the general, or generic, name of the product itself.

Directions

The 10 items below are easy to recognize. Some of them are registered trade names while others merely describe a type of product. Using a dictionary and any other references you wish, determine whether the word is a product name or a trade name. If it is a trade name, record it below and answer the questions that follow. If it is the name of a product, record it on the lines provided following the information about trade names.

KLEENEX	COKE
BAND-AID	VACUUM
DOUGHNUT	ASPIRIN
LAWN MOWER	PERSONAL COMPUTER
XEROX	POLAROID

TRADE NAME _____

What is the generic name for this product? _____

Name three competing products:

_____

_____

_____

What does the company trademark look like?

_____

TRADE NAME _____

What is the generic name for this product? _____

Name three competing products:

_____

_____

_____

What does the company trademark look like?

_____

TRADE NAME _____

What is the generic name for this product? _____

Name three competing products:

_____

_____

_____

What does the company trademark look like?

_____

TRADE NAME _____

What is the generic name for this product? _____

Name three competing products:

_____

_____

_____

What does the company trademark look like?

_____

TRADE NAME _____

What is the generic name for this product? _____

Name three competing products:

_____

_____

_____

What does the company trademark look like?

_____

PRODUCT NAMES:

1. _____   3. _____   5. _____

2. _____   4. _____

WRITING SKILLS— creating an advertising program

Planet Ads, Interrestrial, has just landed a new advertising contract. Five small asteroids have become available for planet-sized advertising. The salespeople at Planet Ads want to make sure they use the asteroids to their best advantage. They have five clients, each of whom sells one of the five products that you listed at the end of the previous exercise. Planet Ads has assigned you the task of deciding which ad will go on which asteroid.

Directions

Place the number of each product name listed at the end of the previous exercise in the crater on the asteroid that would best serve the purposes of the advertisement. Then, using a separate sheet of paper, design a trade name, a trademark, and a billboard advertisement for one of the products above. The following outline should help.

Product Name _____   Your Trade Name _____

Audience that you are trying to reach: _____

_____

What message do you want your billboard illustration to give to potential customers?

_____

_____

Trademark:

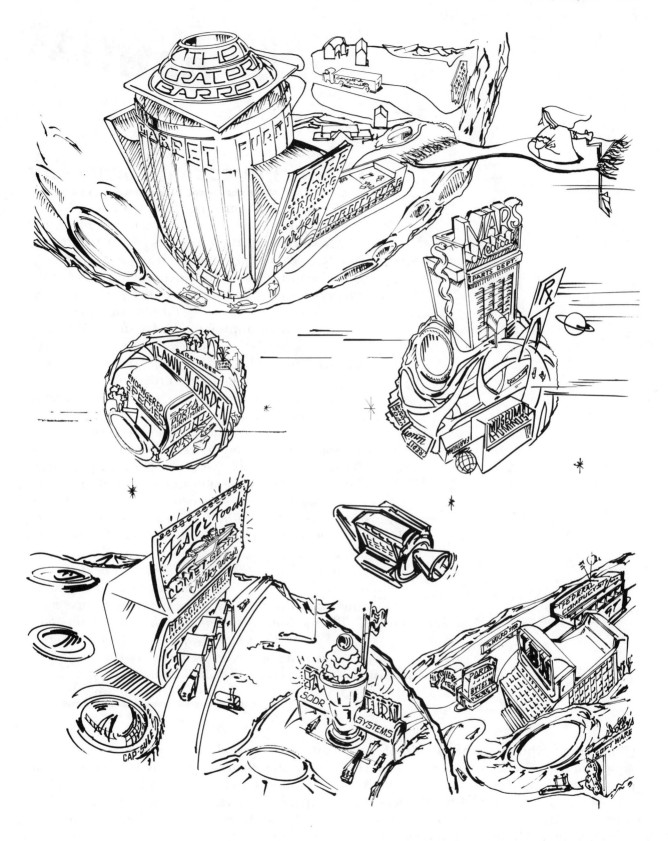

GANTLET
Richard E. Peck

Prereading Vocabulary

fetid: foul smelling, rotting. *(The air in the station became* fetid *as everyone crowded onto the platform.)*

rivulet: a small stream. *(Being in the driver's seat made Brens nervous.* Rivulet(s) *of perspiration covered his forehead.)*

feign: to pretend; to act falsely. *(Many of the commuters preferred to* feign *sleep rather than look out the window at the city.)*

acrid: an unpleasant taste or smell. *(The* acrid *smoke hung like a blanket over the city.)*

subterranean: underground. *(The train followed a* subterranean *route out of the city.)*

clamoring: making a loud noise or disturbance often created by a large crowd.) *(The children began* clamoring *around the oxygen salesmen.)*

statistics: data, numerical information. *(The only thing Brens knew about the people in Opensky came from the* statistics *he read at his office.)*

paradox: a true statement that for some reason seems false. *(To Brens, it seemed that the faster the train traveled, the less movement he felt. He could not explain this* paradox.*)*

paroxysm: a sudden outburst, such as coughing or laughing. *(Brens watched as the man on the tracks bent forward in a* paroxysm *of coughing.)*

malcontent: one who is not happy with his or her present condition. *(The* malcontent(s) *were foolish to complain about their living conditions, Brens thought.)*

anarchist: one who believes that all forms of government should be abolished. *(The* anarchist(s) *might be planning to destroy the city.)*

Jack Brens thumbed the ID sensor and waited for the sealed car doors to open. He had stayed too long in his office, hoping to avoid any conversation with the other commuters, and had been forced to trot through the fetid station. The doors split open; he put his head in and sucked gratefully at the cool air inside, then scrubbed his moist palms along his thighs and stepped quickly into the car. Rivulets of sweat ran down the small of his back. He stretched his lips into the parody of a confident smile.

Most of the passengers sat strapped in, a few feigning sleep, others trying to concentrate on the stiff-dried factsheets which rattled in their hands. Lances of light fell diagonally through the gloom; some of the boiler plate welded over the windows had apparently cracked under the twice-daily barrage.

Brens bit the tip of his tongue to remind himself to call Co-op Maintenance when he got home. Today the train was his responsibility— one day out of one hundred; one day out of twenty work weeks. If he didn't correct the flaws he noticed, he might suffer because of them tomorrow, though the responsibility would by then have shifted to someone else. To whom? Karras. Tomorrow Karras had window seat.

Brens nodded to several of the gray-haired passengers who greeted him.

"Hey, Brens. How's it going?"

"Hello, Mr. Brens."

"Go get 'em, Jack."

He strode down the aisle through the aura of acrid fear rising from the ninety-odd men huddled in their seats. A few of the commuters had already pulled their individual smoking bells down from the overhead rack. Although the rules forbade smoking till the train got underway, Brens understood their feelings too well to make a point of it.

Only Karras sat at the front. The seats beside and behind him were empty.

"Thought you weren't coming and I might have to take her out myself," Karras said. "But my turn tomorrow."

Brens nodded and slipped into the engineer's seat. While he familiarized himself with the instrument console, he felt Karras peering avidly past him at the window. Lights in the station tunnel faded and the darkness outside made the window a temporary mirror. Brens glanced at it once to see the split image of Karras reflected in the inner and outer layers of the bulletproof glass: four bulging eyes, a pair of glistening bald scalps wobbling in and out of focus.

The start buzzer sounded.

He checked the interior mirror. Only two empty seats, at the front of course. He'd heard of no resignations from the Co-op and therefore assumed that the men who should have occupied those seats were ill; it took something serious to make a man miss his scheduled car and incur the fine of a full day's salary.

The train thrummed to life. Lights flared, the fans whined toward full thrust, and the car danced unsteadily forward as it climbed onto its cushion of air. Brens concentrated on keeping his hovering hands near the throttle override.

"You really sweat this thing, don't you?" Karras said. "Relax. You've got nothing to do but enjoy the view, unless you think you're really playing engineer."

Brens tried to ignore him. It was true that the train was almost totally automatic. Yet the man who drew window seat did have certain responsibilities, functions to perform, and no time to waste. No time until the train was safely beyond the third circle—past Cityend, past Opensky, past Workring. And after that, an easy thirty miles home.

Brens pictured the city above them as the train bored its way through the subterranean darkness, pushing it back with a fan of brilliant light. City stretched for thirty blocks from center in this direction and then met the wall of defenses separating it from Opensky. The whole area of City was unified now, finally—buildings joined and sealed against the filth of the air outside that massive, nearly self-sufficient hive. Escalators up and down, beltways back and forth, interior temperature and pollution kept at an acceptable level—it was all rather pleasant.

It was heaven, compared to Opensky. Surrounding and continually threatening City lay the ring of Opensky and its incredible masses of people. Brens hadn't been there for years, not since driving through on his way to work had become impossibly time-consuming and dangerous. Twenty years ago he had been one of the last lucky ones, picked out by Welfare Control as "salvageable"; these days, no one left Opensky. For that matter, no one with any common sense entered.

He could vaguely recall seeing single-family dwellings there, whether his wife, Hazel, believed that claim or not, and more vividly the single-family room he had shared with his parents and grandfather. He could even remember the first O-peddlers to appear on Sheridan Street. Huge, brawny men with green O-tanks strapped to their backs, they joked with the clamoring children who tugged at their sleeves and tried to beg a lungful of straight O for the high it was rumored to induce. But the peddlers dealt at first only with asthmatics and early-stage emphysemics who gathered on muggy afternoons to suck their metered dollar's worth from the grimy rubber face mask looped over the peddler's arm. All that was before each family had a private bubble hooked directly to the City metering system.

He had no idea what life in Opensky was like now, except what he could gather from the statistics that crossed his desk in Welfare Control. Those figures meant little enough: so many schools to maintain, dole centers to keep stocked and guarded, restraint aides needed for various playgrounds—he merely converted City budget figures to percentages corresponding to the requests of fieldmen in Opensky. And he hadn't spoken to a fieldman in nearly a year. But he assumed it couldn't be pleasant there. Welfare Control had recently disbanded and reassigned to wall duty all Riot Suppression teams; the object now was not to suppress, but to contain. What went on in Opensky was the skyers' own business, so long as they didn't try to enter City.

So. Six miles through Opensky to Workring, three miles of Workring itself, where the skyers kept the furnaces bellowing and City industry alive. But that part of the trip wouldn't be too bad. Only responsible skyers were allowed to enter Workring, and most stuck to their jobs for fear of having their thumbprints erased from the sensors at each Opensky exit gate. Such strict control had seemed harsh, at first, but Brens now knew it to be necessary. Rampant sabotage in Workring had made it so. The skyers who chose to work had nearly free access to and from Workring. And those who chose not to work— well, that was their choice. They could occupy themselves somehow. Each year Welfare Control authorized more and more playgrounds in Opensky, and the public schools were open to anyone under fifty with no worse than a moderate arrest record.

Beyond Workring lay the commuter residential area. A few miles of high-rise suburbs, for secretaries and apprentice managerial staff, merging suddenly with the sprawling re-development apartment blocks, and then real country. To Brens the commuter line seemed a barometer of social responsibility: the greater one's worth to City, the farther away he could afford to live. Brens and his wife had moved for

the last time only a year ago, to the end of the trainpad, thirty miles out. They had a small square of yellowed grass and two dwarf apple trees that would not bear. It was ...

He shook off his daydreaming and tried to focus on the darkness rushing toward them. As their speed increased, he paradoxically lost the sense of motion conveyed by the lurching start and lumbering underground passage. Greater speed increased the amount of compression below as air entered the train's howling scoops and whooshed through the ducts down the car sides. Cityend lay moments ahead.

Brens concentrated on one of the few tasks not yet automated: at Cityend, and on the train's emergence from the tunnel, his real duty would begin. Three times in the past month skyers had sought to breach City defenses through the tunnel itself.

"Hey! You didn't check defense systems," Karras said.

"Thanks," Brens muttered through clenched teeth. "But they're okay." Then, because he knew Karras was right, he flipped the arming switch for the roof-mounted fifties and checked diverted-power availability for the nose lasers. The dials read in the green, as always.

Only Karras, who now sat hunched forward in anticipation, would have noticed the omission. Because Karras was sick. The man actually seemed to look forward to his turn in the window seat, not only for the sights all the other commuters in the Co-op tried to avoid, but also for the possible opportunity of turning loose the train's newly installed firepower.

"One of these days they're going to make a big try. They'd all give an arm to break into City, just to camp in the corridors. Now, if it was me out there, I'd be figuring a way to get out into Suburbs. But them? All they know is destroy. Besides, you think they'll take it lying down that we raised the O-tax? Forget it! They're out there waiting, and we both know it. That's why you ought to check all the gear we've got. Never know when ..."

"Later, Karras! There it is." Brens felt his chest tighten as the distant circle of light swept toward them—tunnel exit, Cityend. His forearms tensed and he glared at the instruments, waiting for the possibility that he might have to override the controls and slam the train to a stop. But a green light flashed; ahead, the circle of sky brightened as the approaching train tripped the switch that cut off the spray of mist at the tunnel exit. And with that mist fading, the barrier of twenty thousand volts which ordinarily crackled between the exit uprights faded also. For the next few moments, while the train snaked its way into Opensky, City was potentially vulnerable.

Brens stared even harder at the opening, but saw nothing. The car flashed out into gray twilight, and he relaxed. But instinct, or a random impulse, drew his eyes to the train's exterior mirrors. And then he saw them: a shapeless huddle of bodies pouring into the tunnel back toward City. He hit a series of studs on the console and braced himself for the jolt.

There it was.

A murmur swept the crowded car behind him, but he ignored it and stared straight ahead.

"What the hell was it?" Karras asked. "I didn't see a thing."

"Skyers. They were waiting, I guess till the first car passed. They must have figured no one would see them that way."

"I don't mean who. I mean, what did you use? I didn't hear the fifties."

"For a man who's taking the run tomorrow, you don't keep up very well. Nothing fancy, none of the noise and flash some people get their kicks from. I just popped speedbreaks on the last three cars."

"In the tunnel? My God! Must have wiped them all the way out the tunnel walls, like a squeegee. Who figured that one?"

"This morning's Co-op bulletin suggested it, remember?"

Karras sulked. "I've got better things to do than pay attention to every word those guys put out. They must spend all day dictating memos. We got a real bunch of clods running things this quarter."

"Why don't you volunteer?"

"I give them my four days' pay a month. Who needs that mishmash?"

Brens silently agreed. No one enjoyed keeping the Co-op alive. No one really knew how. And that was one of the major problems associated with having amateurs in charge: it's a hell of a way to run a railroad. But the only way, since the line itself had declared bankruptcy, and both

city and state governments refused to take over. If it hadn't been for the Co-op, City would have died, a festering ulcer in the midst of the cancer of Opensky.

Opensky whirled past them now. Along the embankment on both sides, legs dangled a decorative fringe. People sat atop the pilings and hurled debris at the speeding stainless steel cars. Their accuracy had always amazed Brens. Even as he willed himself rigid, he flinched at the eggs, rocks, bottles, and assorted garbage that clattered and smeared across the window.

"Look at those sonsabitches throw, would you? You ever try and figure what kind of lead time you need to hit something moving as fast as we are?"

Brens shook his head. "I guess they're used to it."

"Why not? What else they got to do but practice?"

Behind them, gunfire crackled and bullets pattered along the boiler plate. Many of the commuters ducked at the opening burst.

"Look at them back there." Karras pointed down the aisle. "Scared blue, every one of them. I know this psychologist who's got a way to calm things down, he says. He had this idea to paint bull's-eyes on the sides of the cars, below the window. Did I tell you about it? He figures it'll work two or three ways. One, if the snipers hit the bull's-eyes, there's less chance of somebody getting tagged through a crack in the boiler plate. Two, maybe they'll quit firing at all, when they see we don't give a suck of sky about it Or three, he says, even if they keep it up, it gives them something to do, sort of channels their aggression. If they take it out on the trains, maybe they'll ease up on City. What do you think?"

"Wouldn't it make more sense to put up shooting galleries in all the playgrounds? Or figure a way to get new cars for the trains? We can't keep patching and jury-rigging these old crates forever. The last thing we need right now is to make us more of a target than we already are."

"Okay. Have it your way. Only, I was thinking . . ."

Brens tuned him out and squinted at the last molten sliver of setting sun. Its rays smeared rainbows through the streaked eggs washing slowly across the window in the slipstream. The mess coagulated and darkened as airblown particles of ash settled in it and crusted over. When he could stand it no longer, Brens flipped on the wipers and watched the clotted slime smear across the glass, as he had known it would. But some of it scrubbed loose to flip back alongside the speeding train.

The people were still out there. If he looked carefully straight ahead, their presence became a mere shadow at the edges of the channel through which he watched the trainpad reeling toward him. Though he doubted any eye would catch his long enough to matter, he avoided the faces. There was always the slight chance that he might recognize one of them. Twenty years wasn't so long a time. Twenty years ago he had watched the trains from an embankment like these.

Now the train swooped upward to ride its cushion of air along the raised pad, level with second-story windows on each side. Blurred faces stared from those windows, here disembodied, there resting on a cupped hand and arm propped on a window ledge. The exterior mirrors showed him faces ducking away from the gust of wind fanning out behind the train and from the debris lifted whirling in the grimy evening air. He tried to picture the pattern left by the train's passage—dust settling out of the whirlwind like the lines of polarization around a magnet tip. A few of the faces wore respirators or simple, and relatively useless, cotton masks. Many didn't bother to draw back but hung exposed to the breeze that the train was stirring up. And now, as on each of his previous rare turns at the window seat, Brens had the impulse to slow the train, to let the wind die down and diminish behind them, out of what he himself considered misplaced and maudlin sympathy for the skyers, who seemed to enjoy the excitement of the train's glistening passage. It tempered the boredom of their day.

"... right about here the six-thirty had the explosion. Five months ago. Remember?"

"What?"

"Explosion. Some kids must have got hold of detonator caps and strung them on wires swinging from a tree. When the train hit them, they cracked the window all to hell. Nearly hurt somebody. But the crews came out and burned

down all the trees along the right of way. Little bastards won't pull *that* one again."

Brens nodded. There was one of the armored repair vans ahead, on a siding under the protective stone lip of the embankment.

The train rose even higher to cross the river which marked the Opensky-Workring boundary. They were riding securely in the concave shell of the bridge. On the river below, a cat, or dog—it was hard to tell at this distance—picked its cautious way across the crusted algae which nearly covered the stream. The center of the turgid river steamed a molten beige; and upriver a short way, brilliant patches of green marked the mouth of the main Workring spillway.

At the far end of the bridge, a group of children scrambled out of the trough of the trainbed to hang over the side.

"Hey! Hit the lasers. Singe their butts for them." Karras bounced in his seat.

"Shut up for a minute, can't you? They're out of the way."

"Now what's that for? Can't you take a joke? Besides, you know they're sneaking into Workring to steal something. You saying we ought to let them get away with it?"

"I'm just telling you to shut up. I'm tired, that's all. Leave it at that."

"Sure. Big deal. Tired! But tomorrow the window seat's mine. So don't come sucking around for a look then, understand?"

"It's a promise."

Sulfurous clouds hung in the air, and Brens checked the car's interior pollution level. It was a safe 18, as he might have guessed. But the sight of buildings tarnished green, of bricks flaking and molting on every factory wall, always depressed him. The ride home was worse than the trip into City. Permissive hours ran from five to eight, when pollution controls were lifted. He knew the theory: evening air was more susceptible to condensation because of the temperature drop, and dumping pollutants into the night sky might actually bring on a cleansing rain. He also knew the practical considerations involved: twenty-four-hour control would almost certainly drive industry away. Compromise was essential, if City was to survive.

It would be good to get home.

The train swung into its gently curving

descent toward Workring exit, and Brens instinctively clasped the seat arms as the seat pivoted on its gimbals. At the foot of the curve he saw the barricade. Something piled on the pad.

Not for an instant did he doubt what he saw. He lunged at the power override, but stopped himself in time. Dropping to the pad now, in mid-curve, might tip the train or let it slide off the pad onto the potholed and eroded right of way where the uneven terrain offered no stable lift base for getting underway again.

"Ahead of you! On the tracks!" Karras reached for the controls, but Brens caught him with a straight-arm and slammed him to the floor. He concentrated on the roadbed flashing toward them. At the last instant, as the curve modified and tilted toward level, he popped all speedbreaks and snatched the main circuit breaker loose.

From the sides of the cars vertical panels hissed out on their hydraulic pushrods to form baffles against the slipstream, and the train slammed to the pad. Tractor gear whined in protest, the shriek nearly drowning out the dying whirr of compressor fans, and the train shuddered to a stop.

Inside, lights dimmed and flickered. Voices rose in the darkness amid the noise of men struggling to their feet.

Brens depressed the circuit breaker and hit the emergency call switch overhead. "Hold it!" he shouted. "Quiet down, please! There's something on the pad, and I had to stop. Just keep calm. I've signaled for the work crews, and they'll be here any minute."

Then he ignored the passengers and focused his attention on the window. The barricade lay no more than twenty feet ahead, rusted castings and discarded mold shells heaped on the roadbed. The jumbled pile seemed ablaze in the flickering red light from the emergency beacons rotating atop the train cars. Behind the barricade and along the right of way, faceless huddled forms rose erect in the demonic light and stood motionless, simply staring at the train. The stroboscopic light sweeping over them made each face a swarm of moving, melting shadows. Brens fired a preliminary burst from the fifties atop the first car, then quickly switched them to automatic, but the watching forms stood like statues.

"They must know," Karras said. He stood beside Brens and massaged his bruised shoulder. "Look. None of them moving."

Then one of the watchers broke and charged toward the car, waving a club. He managed two strides before the fifties homed on his movement and opened up. A quick chatter from overhead and the man collapsed. He hurled the club as he dropped and the fifties efficiently followed its arc through the air with homed fire that made it dance in a shower of flashing sparks. It splintered to shreds before it hit the ground.

The other watchers stood motionless.

Brens stared at them a long moment before he could define what puzzled him about their appearance: none of them wore respirators. Were they trying to commit suicide? And why this useless attack? His eyes had grown accustomed to the flickering light and he scanned the mob. Young faces and old, mostly men but a few women scattered among them, all shades of color, united in appearance only by their clothing. Workring skyers in leather aprons, thick-soled shoes, probably escapees from a nearby factory. He flinched as one of them nodded slightly—surely they couldn't see him through the window. The nod grew more violent, and then he realized that the man was coughing. Paroxysms seized the man as he threw his hands to his mouth and bent forward helplessly. It was enough. The fifties chattered once more, and he fell.

"But what do they get out of it?" He turned his bewilderment to Karras.

"Who can tell? They're nuts, all of them. Malcontents, or anarchists. Mainly stupid, I'd say. Like the way they try and break into City. Even if they threw us out, they wouldn't know what to do next. Picture one of them sitting in your office. At your desk."

"I don't mean that. If they stop us from getting through, who takes care of them? I mean, we feed them, run their schools, bury them. I don't understand what they think all this will accomplish."

"Listen! The crew's coming. They'll take care of them."

A siren keened its rise and fall from the dimming twilight ahead, but still the watchers stood frozen. When the siren changed to a blatting klaxon, Brens switched the fifties back on manual to safeguard the approaching repair car. The mob melted away at the same signal. They were there, and then they were gone. They dropped from sight along the pad edge and blended into the shadows.

The work crew's crane hoisted the castings off the pad and dropped them on the right of way. In a few minutes they had finished. Green lights flashed at Brens, and the repair van sped away again.

Passing the Workring exit guards, Brens made a mental note to warn the Co-op. If the skyers were growing bold enough to show open rebellion within the security of Workring, the exit guards had better be augmented. Even Suburbs might not be safe any longer. At thirty miles distance, he wasn't really concerned for his own home, but some of the commuters lived dangerously close to Workring.

He watched in the exterior mirror. The rear car detached itself and swung out onto a siding where it dropped to a halt while the body of the train went on. Every two miles, the scene repeated itself. Cars dropped off singly to await morning reassembly. Brens had often felt a strange sort of envy for the commuters who lived closer in: they never had the lead window seat on the way out of City. Responsibility for the whole train devolved on them only for short stretches, only on the way in.

But that was fair, he reminded himself. He lived the farthest out. With privilege go obligations. And he was through, for another twenty weeks, his obligations met.

At the station, he telexed his report to the Co-op office and trotted out to meet Hazel. The other wives had driven away. Only his carryall sat idling at the platform edge. He knew he ought to look forward to relaxing at home, but the trip itself still preyed on his mind unaccountably. He felt irritation at his inability to put the skyers out of his thoughts. His whole day was spent working for their benefit; his evenings ought to be his own.

He looked back toward City, but saw nothing in the smog-covered bowl at the foot of the hills that stretched away to the east. If it rained tonight, it might clear the air.

Hazel smiled and waved.

He grinned in answer. He could predict her reaction when she heard what he'd been through: a touch of wifely fear and concern for him, and that always made her more affectionate. Almost a hero's welcome. After all, he had acquitted himself rather well. A safe arrival, only a few minutes late, no injuries or major problems. And he wouldn't draw window seat for another several months. It was good to be home.

POSTREADING VOCABULARY EXERCISE

Directions

Select the best word, or form of the word, and write it on the appropriate line below. Refer to the Prereading Vocabulary for help if needed.

1. feigned—fetid

   The sanitation workers _____ illness to avoid picking up

   the _____ garbage that collected after the flood.

2. acrid—rivulets

   The _____ steam rose from the

   _____ of acid flowing over the barren land.

3. statistics—subterranean

   There are few _____ available regarding the number of

   _____ rail accidents.

4. paradox—paroxysms

   The doctors were unable to explain the _____ of an otherwise

   healthy man suffering with continuing _____ of wheezing.

5. anarchists—malcontents—clamoring

   The _____ began _____

   for the downfall of the government while the _____ called for
   sweeping changes.

THINKING ABOUT WHAT YOU'VE READ

1. Society had become as divided as the physical barriers that surrounded City. Why were the people in the suburbs so afraid of the people of Opensky?
2. Why was the railroad abandoned by its owners? Why was it now run by the passengers?
3. Karras enjoyed taking the controls of the train while Brens hated the job. Knowing only this about them, how would you describe the two men?

COMPREHENSION— author's purpose

Directions

Once you begin reading a story, you lend yourself to the author. It is through the author's writing that you sense all that happens to the characters in the story. The author can give you as much — or as little — information as he or she wishes and can use a variety of techniques to achieve a specific purpose, such as introducing a particular character, changing the reader's perspective, or changing the pace of the action. Answer the following questions about the author's purpose in the story.

1. What was author Peck's purpose in describing Brens' moist palms, sweating back, and feigned smile in the first paragraph?

_____

_____

2. What impression of Brens does the author give you in the first paragraph? (Check one.)
   ☐ confident   ☐ nervous   ☐ tired   ☐ happy

3. In the second paragraph, what does the author do to let the reader know that this is not an ordinary train?

_____

_____

4. What visual image of the commuter train do you form after reading the second paragraph?

_____

_____

5. The author introduces the character of Karras after mentioning the commuters who were smoking. Describe your impression of Karras after reading the next two paragraphs.

_____

_____

6. Why do you think the author leads you to form this impression of Karras?

_____

_____

COMPREHENSION— distinguishing between fact and opinion

Directions

The trip to the suburbs took the commuters through several population rings. What those on the train knew about the inhabitants of the city rings came from two sources, the statements by Welfare Control and the opinions of their fellow commuters. Read the following statements. If a statement expresses an opinion, place an **O** on the line preceding the number. If it states a fact, place an **F** on the line.

_____ 1. The air in Opensky was so polluted that the city supplied each household with a direct oxygen line.

_____ 2. No one was ever allowed to leave Opensky.

_____ 3. Nobody with any sense entered Opensky.

_____ 4. Living in the city was heaven compared with conditions in Opensky.

_____ 5. People attempted to leave Opensky by trying to escape through the train tunnels.

_____ 6. The commuter who rode in the only seat near a window had the responsibility of guiding and protecting the train during the trip to the suburbs.

_____ 7. Riding in the window seat was considered to be a terrible obligation.

_____ 8. The public schools in Opensky were open to anyone under 50 with no worse than a moderate arrest record.

_____ 9. The skyers were all dedicated to taking over the city.

_____ 10. The skyers were bold enough to barricade the train tracks.

RESEARCH SKILLS— writing a social history

Brens, in his job with Welfare Control, saw a variety of statistics about life in the population rings surrounding the city. While he had a vague idea of how many people were living in Opensky and the type of housing they had, he didn't have as much information about how the skyers lived — what they did with their days and nights.

Social history looks at more than just statistics. It looks at the quality of lives, not just the quantity. Among other things, it examines how people live, their entertainment, leisure activities, education, family, and transportation.

Directions

Pick a country and a time period from the first two columns below and pair them with one of the topics in the third column. Then do the research necessary to write a short social history about the topic in the country and time period that you selected.

COUNTRY	TIME PERIOD	TOPIC
USA (from 1776)	1700–1799	Entertainment
England	1800–1899	Leisure Activities
France	1900–present	Education
		Family
		Transportation

As you collect your information from various publications, you will want to record it on bibliography cards like the one below. The information you record on these cards should include the following:

- The card number and the library card catalog number
- The title of the book or article
- Who wrote the book or article
- The name and city of the publisher
- The publication date

Use the rest of the card to record any interesting quotations you found in the book. These notes are called annotations.

Card Number _____

Library Card Catalog Number _____

Title of Publication _____

Author _____

Publisher _____

City Where Published _____

Year Published _____

Annotations _____

_____

_____

After compiling your bibliography cards, the outline below will help you organize your research. Major points that you learned through your research should be written on the lines with roman numerals; information that supports your major point should be listed on the lines beginning with capital letters.

I. _____

   A. _____

   B. _____

   C. _____

II. _____

   A. _____

   B. _____

   C. _____

III. _____

    A. _____

    B. _____

    C. _____

WRITING SKILLS— writing from an outline

Directions

Using your bibliography cards and outline, on a separate sheet of paper write a social history on the subject you've researched.

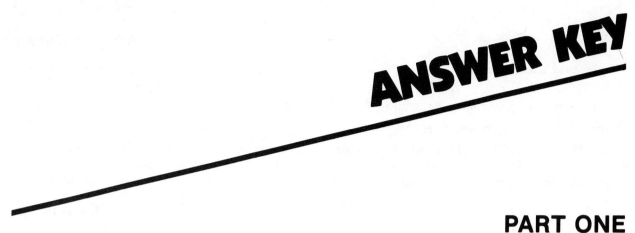

PART ONE

How Now Purple Cow

Postreading Vocabulary Exercise: foreman; oblivious; grazing; chlorophyll; hallucinating; corrugated.

Comprehension—referents: 1. Floyd; Floyd; the cow. 2. the telephone; Jim Player. 3. C; A; B.

Comprehension—recall: 1. X; 2. X; 3. 1; 4. X; 5. 3; 6. X; 7. X; 8. 4; 9. 2; 10. X.

Just Call Me Irish

Postreading Vocabulary Exercise: 1. humorous; 2. owner; 3. scales; 4. recording; 5. shy; 6. sit; 7. canine.

Comprehension—character: 1. surprised; 2. bored; 3. irritated; 4. discriminated against; 5. surprised.

Comprehension—dialogue: 1. abruptly stopped speaking; 2. C; 3. answers will vary; 4. answers will vary; 5. answers will vary; 6. answers will vary; 7. answers will vary.

Zoo

Postreading Vocabulary Exercise: 1. professor; 2. chatter; 3. fascinated; 4. wonderment; 5. offspring; 6. breed; 7. jagged; 8. horrified; 9. San Diego, California.

Comprehension—stating correct sequence: 1. 4; 2. 1; 3. 2; 4. 7; 5. 3; 6. 6; 7. 5.

Comprehension—cause and effect: 1. D; 2. E; 3. B; 4. A; 5. F; 6. C.

Dog Star

Postreading Vocabulary Exercise: 1. B; 2. D; 3. A; 4. E; 5. C.

Comprehension—true/false: 1. F; 2. F; 3. F; 4. T; 5. T; 6. T; 7. T; 8. T; 9. T; 10. T.

Comprehension—assumptions: 1. A, C; 2. B, D; 3. A, B; 4. B, D.

Creature of the Snows

Postreading Vocabulary Exercise: 1. escarpment; 2. plateau; 3. vista; 4. pinnacle; 5. massif; 6. crevass; 7. abyss; 8. grandeur (answers will vary); 9. acclimatized (answers will vary).

Comprehension—supporting details: 1. A. The group had found nothing; B. They had done nothing but climb; C. Dr. Schenk, the expedition leader, argued with the guides. 2. A. He took a picture of the creatures; B. The pilot spotted the creatures playing at an altitude of 25,000 feet; C. He was unable to find an updraft; D. Soaring over Everest was a goal he had not achieved. 3. A. He planned to leave at daybreak; B. He planned to travel light—no equipment except for the necessary oxygen tanks; C. He planned to take just one guide up the mountain with him.

Comprehension—comparison: 1. pilot, photographer; 2. pilot; 3. photographer; 4. pilot, photographer; 5. pilot; 6. pilot.

PART TWO

King of Beasts

Postreading Vocabulary Exercise: 1. C;   2. C;   3. I;   4. I;   5. C;   6. C;   7. C;   all other responses will vary.

Comprehension—cause and effect: all answers will vary.

Comprehension—summary: all answers will vary.

Study Skills—classification: *Prehistoric*—glyptodant, baluchistan, saber-toothed tiger, eohippus, woolly mammoth; *Extinct*—passenger pigeon, dodo, moa, great auk, urus (wild ox): *Endangered*—woolly spider monkey, brown pelican, Japanese sea lion, Florida Key deer, bald eagle.

Little William

Postreading Vocabulary Exercise: 1. A, consciousness; 2. B, spinster; 3. B, inflexible; 4. A, beguile; 5. B, solitude; 6. A, brood; 7. B, warrant; 8. B, unparalleled; 9. A, tremendous.

Comprehension—sequencing: 1. WH, 1;   2. LW, 8;   3. LW, 5;   4. LW, 10;   5. WH, 2;   6. LW, 7;   7. WH, 6;   8. WH, 4;   9. LL, 9;   10. LL, 3.

Comprehension—summarizing: all answers will vary.

Appointment at Noon

Postreading Vocabulary Exercise:

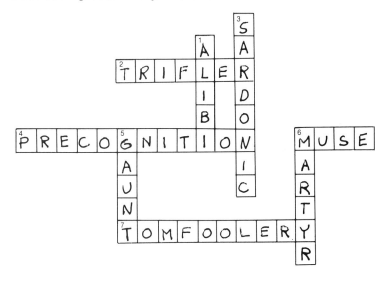

Comprehension—identification/sequence: 1. E (Henry);   2. K (secretary);   3. D (Henry);   4. B (secretary);   5. A (Henry);   6. G (old man);   7. H (secretary);   8. J (Henry);   9. C (old man);   10. L (Henry);   11. F (old man);   12. I (Henry).

Comprhension—figurative language: all answers will vary.

The Boy with Five Fingers

Postreading Vocabulary Exercise: 1. poison; 2. fall; 3. barren; 4. baby; 5. mansion.

Comprehension—comparison: 1. old race, new order; 2. old race; 3. new order; 4. new order; 5. new order.

Comprehension—fact and opinion: 1. F;   2. O;   3. F;   4. O;   5. O;   6. F;   7. O;   8. answers will vary; 9. answers will vary; 10. answers will vary.

The Last Paradox

Postreading Vocabulary Exercise: 1. gesture; 2. portray; 3. vapor; 4. condense; 5. vibrate; 6. occupant; 7. inherent; 8. auxiliary.

Comprehension—supporting details: 1. Comptoss prepared for time travel, delete D; 2. Fordley had spent time developing …, delete A; 3. Time travel was not what people had expected, delete D.

Comprehension—recall: 1. A. disappointed, B. answers will vary; 2. A. answers will vary, B. confused, C. answers will vary; 3. A. answers will vary, B. answers will vary; 4. A. answers will vary, B. answers will vary, C. answers will vary.

PART THREE

Dreamworld

Postreading Vocabulary Exercise: 1. E (devotee); 2. A (pious); 3. H (deceased); 4. B (waver); 5. F (tolerate); 6. G (exasperate); 7. D (severe); 8. C (coalesce).

Comprehension—fantasy/reality: 1. R; 2. F; 3. R; 4. R; 5. F; 6. F; 7. F; 8. R; 9. R; 10. F.

Comprehension—figurative language: all answers will vary.

Kin

Postreading Vocabulary Exercise: 1. A; 2. B; 3. B; 4. A; 5. D; 6. C.

Comprehension—sequence: 1. 4; 2. 2; 3. 6; 4. 1; 5. 3; 6. 5.

Comprehension—fact and opinion: 1. F; 2. O; 3. O; 4. F; 5. F; 6. O; 7. F; 8. F; 9. F; 10. O.

The Fun They Had

Postreading Vocabulary Exercise: 1. E; 2. D; 3. A; 4. B; 5. C.

Comprehension—comparisons: 1. T; 2. G; 3. B; 4. G; 5. G; 6. T; 7. G; 8. T; 9. T.

Comprehension—fact/opinion: 1. F; 2. O; 3. O; 4. F; 5. O; 6. F; 7. O; 8. F; 9. F; 10. answers will vary.

Hometown

Postreading Vocabulary Exercise: 1. perpetual; 2. artificial; 3. indulge; 4. neurotic; 5. authentic.

Comprehension—drawing conclusions: 1. home; 2. church; 3. park; 4. fire station; 5. hospital.

Comprehension—similarities/differences: 1A. Each is constructed of brick and wood; 1B. Unlike the houses on Earth, the Hometown houses have lawns that require little maintenance; 2A. They both spent the entire day walking through Hometown; 2B. She was ready to spend the next six months at Hometown; 2C. He was very tired by the end of the day.

Speed of the Cheetah, Roar of the Lion

Postreading Vocabulary Exercise: 1. conveyance; 2. slipstream; 3. exhaust; 4. ration; 5. coincidence; 6. behemoth; 7. surge; 8. juggernaut; 9. ignition; 10. accelerator.

Comprehension—reality/illusion: 1. R; 2. I; 3. I; 4. R; 5. R.

Comprehension—comparisons: 1. Henry; 2. 1, 2; 3. 1, 2; 4. Henry; 5. Simon; 6. Simon; 7. Simon; 8. answers will vary; 9. because Henry's car was a source of constant irritation.

Buy Jupiter

Postreading Vocabulary Exercise: 1. intrinsic; 2. artificial; 3. idolizing; 4. neutrality; 5. bickering.

Comprehension—forming conclusions: 1. kept blazing hot; 2. in the haloes of O-spectra stars;
3. too weak to support Mizzarett life; *the Mizzaretts had no interest in settling on the Earth;
4. preferred to enter into legal treaties with their neighbors; 5. agreed to supply Earth with energy;
6. the Mizzaretts did not want the Lamberj people to learn of the plan; * the Mizzarett plan was of importance only to traveling Mizzaretts and Lamberj.

Comprehension—predicting outcomes: 1. answers will vary; 2. answers will vary.

Gantlet

Postreading Vocabulary Exercise: 1. feigned, fetid; 2. acrid, rivulets; 3. statistics, subterranean;
4. paradox, paroxysms; 5. anarchists, clamoring, malcontents.

Comprehension—author's purpose: answers will vary.

Comprehension—fact/opinion: 1. F; 2. F; 3. O; 4. O; 5. F; 6. F; 7. O; 8. F; 9. O; 10. F.

INDEX